Forest Under Story

FOREST UNDER STORY

CREATIVE INQUIRY *in an* OLD-GROWTH FOREST

Edited by
NATHANIEL BRODIE
CHARLES GOODRICH
FREDERICK J. SWANSON

A Ruth Kirk Book
UNIVERSITY OF WASHINGTON PRESS
Seattle & London

Forest Under Story is published with the assistance of a grant from the Ruth Kirk Book Fund, which supports publications that inform the general public on the history, natural history, archaeology, and Native cultures of the Pacific Northwest.

Printed and bound in the United States of America
Composed in Cassia, a typeface designed by Dietrich Hofrichter
Design: Dustin Kilgore / Photographs: Bob Keefer
20 19 18 17 16 5 4 3 2 1

UNIVERSITY OF WASHINGTON PRESS
www.washington.edu/uwpress

Cataloging information is on file with the Library of Congress
ISBN 978-0-295-99545-8

The paper used in this publication is acid-free and meets the minimum requirements of American National Standard for Information Sciences—Permanence of Paper for Printed Library Materials, ANSI Z39.48–1984. ∞

To Jim Sedell

1944–2012

“What’s the story?”

He set us on this path.

The moral I labor toward is that a
landscape . . . can best be understood
and given human significance by
poets who have their feet planted
in concrete—concrete data—and by
scientists whose heads and hearts
have not lost the capacity for wonder.

—**Ed Abbey**

Contents

Forest Under Story

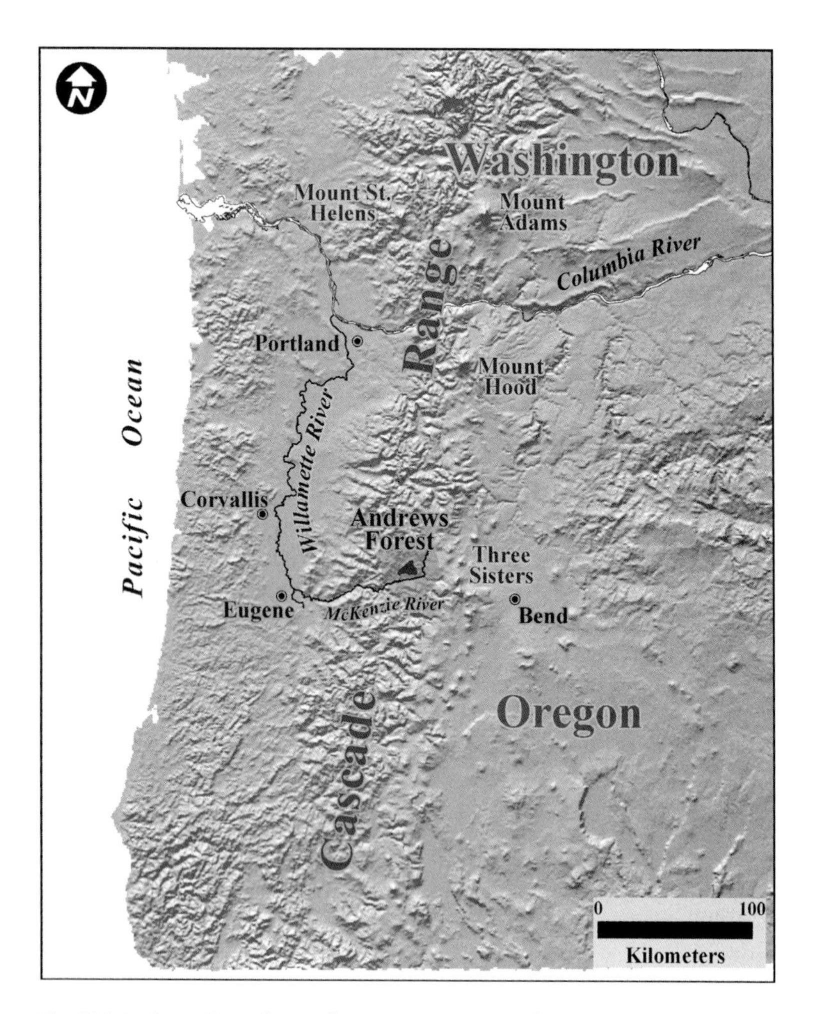

The H. J. Andrews Experimental Forest encompasses the entire Lookout Creek watershed within the Willamette National Forest on the west slope of the Cascade Range, the backbone of Pacific Northwest terrain. Winter storms from the North Pacific drive the climate of cool, wet winters and warm, dry summers, which support forests of massive, long-lived conifers characteristic of the region and well represented in the Andrews Forest.

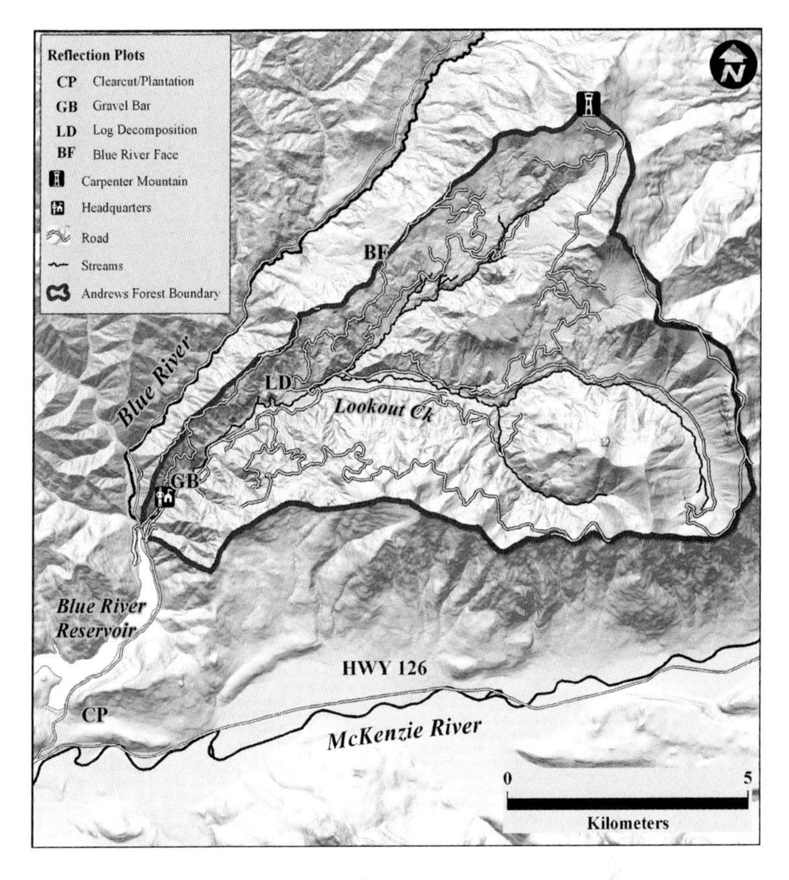

Broadly representative of the rugged mountainous landscape of the Pacific Northwest, the Andrews Forest contains excellent examples of the region's conifer forests and associated wildlife and stream ecosystems. Four Reflections Plots within and near the Andrews Forest are points of entry for visiting writers during their inquiries into the long-term workings of the forest and streams and the cultural landscape.

Entries into the Forest

CHARLES GOODRICH

Take Oregon 126 east out of Eugene, then turn upslope onto a gravel road and wind deep into the Cascade Mountains, and park. Follow a duff-covered path into a grove of old-growth conifers, and stand quietly amid the massive boles. Gaze up into the interlaced limbs of the canopy. With the moss-cushioned ground underfoot and the wavering shafts of afternoon sunlight overhead, you may feel yourself opening into a beautifully benign relationship with the forest. The towering scale, the earthy smells, the subtle sounds, the play of light on moss, lichen, fir needle, and fern will inform your whole being that you are in one of the most vivacious places on Earth. You will understand why such groves have been likened to cathedrals.

On the other hand, if you scramble off trail and bushwhack into the tangled understory, you may enter into a different sort of relationship: a wrestling match with the landscape. The terrain will be so steep you'll have to haul yourself uphill hand-over-hand, grasping at vine maple or rhododendron or devil's club. The understory shrubs will be thick in your face, as will, in early summer, mosquitoes. Or if you've waited until fall to avoid the mosquitoes, it will be raining, or if isn't raining it will have recently been raining, and every leaf and branch will sop your boots, your clothing, your naked face.

In the absence of a trail, the forest can seem claustrophobic, disorienting, even maddening. Unless you have a powerful reason to continue—unless you're a hunter tracking a deer, say, or a timber cruiser marking a sale, or a scientist counting salal along a transect, or a poet searching for the genius of place—you will retreat to the nearest road and look back over your shoulder as if you've survived a mugging. You will feel grateful for the human-made path, for the relatively few and far-between roads that wind through the topographic and vegetative jumble.

Trying to comprehend what these forests *mean* can be just as bewildering. We may intuit and celebrate the wholeness of the forest, but we know it in pieces and threads, by its species and cycles, its products and processes. We come to know the forest via the paths laid down in stories, stories told in anecdotes, photographs, essays, and poems, or in hypotheses, data, and graphs. All these stories are entries into the forest, paths that others have made and which we may follow, perhaps to discover new insights and entice others to enter too.

Throughout most of U.S. history, those entering the forest have done so to hunt, fish, or gather; to harvest timber; or to dig for minerals. Or to raze the forest to convert the land to farms and cities. Forest research is a more recent way of entry into the forest. Since 1948 the entire 15,800-acre watershed of Lookout Creek in the Oregon Cascade Range has been dedicated to the quest, not for lumber or fiber, but for knowledge. The H. J. Andrews Experimental Forest is a storied place among forest scientists. Research conducted here has been instrumental in discovering the unique subecosystems of the forest canopy, the role of dead wood in stream and forest ecosystems, the behavior and ecological role of the spotted owl and other animals, and the nature of old growth itself.

Among the scientists who have worked at the Andrews Forest, some have been especially aware of the importance of framing their efforts in compelling ways. Stream ecologist Jim Sedell, an early proponent for bringing creative writers to the Andrews, was known to ask his fellow researchers, "What's the story here?" It was a conversation between Sedell and writer and philosopher Kathleen Dean Moore that jump-started the creation of the Long-Term Ecological Reflections program.

Since its beginnings in 2003, the Reflections program has made it possible for creative writers—people for whom "What's the story?"

is a primary mode of investigation—to spend one to two weeks in residence at the Andrews Forest. They are invited to pursue their own original inquiries using the methods of the humanities—imagination, metaphor, direct experience, research—to follow paths both literal and metaphorical into the forest. Among the Andrews's ancient, moss-draped trees, its tree plantations after recent clear-cutting, its stream-swept gravel bars, and its hillsides scoured by fire, they are encouraged to walk, observe, reflect, and record their insights.

This book collects some of the best writings from the first dozen years of the Reflections program. In a diversity of forms, including essays, field notes, and poems, by a diverse range of writers—some of whom were trained as scientists, some who are professional writers, some who are both—these cultural data offer a fascinating record of some of the many ways we approach, experience, and understand the forest and the relation between people and the forest. The book, as Robin Kimmerer writes in these pages, is "a chronicle of the land, a witnessing of the world, understanding and wonder," perhaps even "a way to predict our impact on the land." It is, she adds, a way that might even "bring us . . . closer to saving what we love."

THE REFLECTIONS PROGRAM

Long-Term Ecological Reflections (or Reflections for short) is run by the Spring Creek Project for Ideas, Nature, and the Written Word at Oregon State University in collaboration with the Andrews Forest Long-Term Ecological Research group and the U.S. Forest Service. Since 1980 the Andrews Forest has been part of the National Science Foundation's Long-Term Ecological *Research* Network. Reflections is intended to be a humanities analog to the scientific research, fostering original, creative inquiry into the ever-changing biophysical and cultural landscapes and gathering the writings as part of a long-term, cumulative record. Because forests and watersheds, along with our cultural attitudes toward them, change both gradually and abruptly, the Long-Term Ecological Research program and Reflections are founded on a commitment to inquiry into ecological and human change spanning generations. And both programs encourage planned outcomes while acknowledging that the most interesting discoveries often come as surprises.

From its inception, the founders of Reflections hoped it could serve as a model that other programs might choose to adopt—other Long-Term Ecological Research sites, which represent major types of ecosystems across the country, or other kinds of place-based programs, some grounded in science and others grounded more in the arts and humanities. At present, many sites have initiated programs that bring together arts, humanities, and science in differing configurations. Some have taken inspiration from Reflections, while others have arisen independently. The Spring Creek Project maintains a website, www.Ecologicalreflections.com, which features site profiles, contact information, and news of current projects for this growing network of Reflections-type programs.

ATTENTION TO PLACE

Reflections is rooted in the belief that when people pay close attention to specific places, their study of place will reveal broad truths that go beyond that place. (As the New Zealand poet Brian Turner wrote during his residency at the Andrews, "remember / this place in other places.") Whether that attention is focused through the lens of science or of art, whether conveyed in the language of fact or of metaphor, there is wisdom to be gained, for the more we know about the natural world and the place of humans in the world, the greater our insight into how we ought to live our lives.

Like other residency programs, Reflections offers lots of unstructured time that allows writers to work on projects of their choice. But unlike most programs, Reflections also asks that each writer enter into the field to visit four sites in the forest, the "Reflections Plots"—a gravel bar created by a recent, major flood, the

Log Decomposition Plot, a recent clear-cut, and an experimental, selectively logged site.

The Andrews contains an intriguing variety of landscapes, from deep forest to open meadows and steep mountain streams. Also evident in the landscape are stream gauges, instrument towers, and the plastic flagging and homemade scientific apparatus cordially referred to as "researcher trash." In fact, data constantly stream in from all over the forest in real time. With fiber-optic cables in the streams, and other remote sensing devices throughout the forest, researchers can gather data on hydrology, soundscape ecology, air flow dynamics, and other ecological processes 24/7, twelve months of the year. It's a place where intense natural processes and fruitful human inquiry are conspicuously intertwined.

Therefore, as crucial as designated wilderness areas are to maintaining cultural and ecological resilience, this book is not a paean to wilderness or untouched nature. Instead, the writers who visit the Andrews Forest engage the many, complicated ways that humans alter nature, whether through forestry management or through the interventions of the science itself. They delve into the ambiguities between utilitarian and intrinsic values, and the paradoxes between ecological preservation and active management of a landscape. It's our belief that creative writers and those who undertake other types of arts- and humanities-based inquires can be especially adept at bringing scientific information and insights into conversation with the complex emotional and cultural relationships that humans have with both wild and managed landscapes.

TAKING THE LONG VIEW

Many ecological changes—the decomposition of fallen trees, the shaping and reshaping of a riverbed, or the reestablishment of a forest after severe fire—involve long-term processes. Likewise, our understanding of the place of humans in the natural world deepens and unfolds over time. That's why the Reflections program is intended to continue for two hundred years. Two centuries—roughly seven human generations—is about the length of time it takes for a big downed log to be transformed back into soil. It offers an ample window of time for chronicling the sweep of change in how people understand a forest ecosystem and their place in it. The long view helps remind us that we can never find a permanent solution or conclusion to any challenge, because the societal context and even the environmental context continuously change—the playing field keeps tilting into the future. Our work in the sciences and humanities will in no way be completed in two hundred years, of course, but we will have learned a great deal from starting with that view.

The aspiration itself demands that we imagine and enact strategies to help assure the continuity of the project. The program's directors are motivated to strengthen its associations with long-lived institutions and, perhaps even more crucially, to cultivate a community of interest that will promote and defend the importance of long-term inquiry in the humanities as well as in the sciences.

Many of the writers in residence at the Andrews have been moved by taking the long-term perspective. The program's first writer in residence, writer and lepidopterist Robert Michael Pyle, emphasized the importance of a commitment to "the long haul." Taking the long view in ecological research and reflection, Pyle writes, requires "faith in the future—even if you won't be there to see it for yourself." Following her residency, writer Alison Hawthorne Deming wrote, "Two hundred years. . . . When I mention this time frame to friends and colleagues, they startle, as if they assume that in two hundred years forests will be either irrelevant or nonexistent. The hope of this project is that by careful and sustained observation, a testimony on behalf of the forest will have kept it alive."

In our experience, commitment to the long view foregrounds the inevitability of ecological change. We become more watchful of our methods, more cautious about our conclusions, more respectful of the integral workings of ecological and biocultural communities, and more attuned to both the presence of the past in the living forest and watershed and the seriousness of our responsibility for shaping the future. And many of the resident writers have suggested that regarding the world across long time scales is itself a source of hope. As Pyle writes, "Maybe looking to the future is a way of hoping there will still be something to see when we get there. Maybe it's the only way to make sure of it."

MULTIPLE WAYS OF KNOWING

Storytelling and poetry, observation and experiment, myth and mathematics are all authentic windows on the world. There is an unusual richness and joy in the community of the humanities and sciences, in the coming together of insights from many different perspectives and disciplines. Science can deeply and significantly enrich the work of creative writers, artists, and philosophers. Reciprocally, we need creative writing, philosophy, the visual arts, and the other humanities to deepen and enrich science. Creative use of language, concepts, and metaphors shape what we can see and imagine. Fresh language and original metaphors can allow us to ask novel questions, conceive new ideas, propose innovative solutions, and bring the experienced world more vividly into the presence of others. Creative writers draw on the rich vocabulary and conceptual insights of science to help people understand and value the world. Scientists can learn to better communicate their ideas in stories that can become part of people's lives. In an effort to heal a damaged landscape, for example, science can recommend tree species to plant and strategies for water-quality improvements, but the causes of habitat degradation reside in the stories people tell themselves and others about their relationship with other creatures, with the processes of nature, and with the land. The long-term success of habitat restoration may depend on a "re-story-ing" of a community's relationship to its landscape by a process that weaves scientific and artistic elements into a compelling narrative for guiding cultural behavior.

Both research and humanities benefit by the sharing, interweaving, or "cross-fertilization" of ideas. As conservation biologist and writer Gary Paul Nabhan has observed, "Science, in and of itself, is seldom enough to reshape public opinion. People have to feel some visceral connection to an issue to act upon it. . . . Artists and scientists . . . need cross-fertilization or else their isolated endeavors will atrophy, wither, or fall short of their aspirations." In rare cases this merging of artistic and technically grounded abilities occurs in individuals, people such as Rachel Carson, John Muir, and Aldo Leopold, who have had a profound impact on thought and action. Since relatively few people embody such multifaceted gifts, perhaps sustained, collective efforts among scientists, writers, and others can offer more than either can separately.

To these ends, Reflections encourages and facilitates opportunities for the resident writers to go into the field with Andrews researchers, offering the writers firsthand encounters with the science and introducing the scientists to the curiosity and insights of the writers. Each writer is taken on a guided, introductory field trip to the Reflections Plots and alerted to the extensive plot descriptions and research information on the Andrews Forest website. At the same time, we remind each writer that we are looking not for science journalism but for original, creative inquiry in the humanities. There are certain kinds of values, such as beauty, surprise, awe, and humility, that both creative writers and scientists may experience and take inspiration from, but that are for the most part proscribed from scientists' professional communications. Creative writers may be able to give these values fuller expression. After Alison Hawthorne Deming went into the field with the spotted owl research crew, she wrote, "Beauty is what I came here for, a beauty enhanced, not diminished, by science. If I had only my senses to work with, how much thinner would be the experience. What a record we might have of the world's hidden beauty if field scientists and poets routinely spent time in one another's company."

The goal in operating Reflections is to foster an ongoing dialogue between the place and the individual writers, and between the science and writers' creative, imaginative responses. Assuming that most readers of *Forest Under Story* will not have visited the Andrews Forest in person or encountered the specific scientific and ecological circumstances that the writers in residence experience there, we've included short "Ground Work" essays by the volume editors in order to present some of the details of the landscape and the research context that our writers in residence encounter on the ground at the Andrews.

* * *

A critical aspect of scientific research at the Andrews Forest is to collect, archive, and actively share data and reports and to encourage critical overviews. Inspired by that commitment, we have from the beginning of the Reflections program been diligent in collecting, archiving, and making available through the program's website, and through its journal, the *Forest Log,* the writings contributed by our residents. *Forest Under Story* continues the effort to share this fund of creative cultural artifacts.

The overarching goal of Reflections is to contribute to the trove of compelling stories that can help humans create and sustain a just and ecologically resilient relationship with each other, with other-than-human creatures, and with the planet. Science, every bit as much as creative writing, is narrative. This book considers the forest under the guise, and under the gaze, of story, to shed light on the primacy of story in understanding our relationship to the forest. It's a foray into re-story-ation. Much of the original research on old-growth forest, for instance, or on the spotted owl, had no immediate relevance at the time. It was pre-relevant. But just as this research became extraordinarily relevant years later, during the "Forest Wars" of the 1990s, so too, we believe, the insights generated by interactions with the forest and recorded in this book may prove to offer keys to questions for which we don't yet have good answers, even questions we have not yet learned to ask.

We are also mindful of the urgency of our larger task of learning to live on Earth without exhausting the habitats of humans and other creatures. We believe that writings such as those collected here play a crucial role by weaving together direct experience, imagination, scientific insights, and careful consideration of human values to better articulate the relationship between humans and the rest of nature. Perhaps our greatest hope for the Reflections project is that it will find ways, through human inquiries such as these, to give a voice to the land itself.

– Part One –

RESEARCH AND REVELATION

The Long Haul

ROBERT MICHAEL PYLE

In the dim deepwood of massive and moss-bound trees, the three tenors of the Northwest forest give voice: varied thrush's raspy note, like whistling through spit; golden-crowned kinglets' high tinkle, the sound older ears lose first; and winter wren, pucks with pennywhistles on an endless tape loop. A fourth, pileated woodpecker, is silent for now, having already totemed all the big old snags.

I've arrived at a place known as the Log Decomposition Plot. The mossy turnoff is paved in evergreen violets, then comes a trench and berm to keep vehicles out, but the bulldozed tank-trap has grown to resemble a native outcrop, covered in sword fern, salal, and moss. Fresh windthrow renders the trail almost impassable at times: a suitable gateway to a place where, when a tree falls in the forest, a lot of people hear it—and then take a close look at what happens next. When I get to the laid-out logs and the sawed-off tree-rounds that fallers call cookies, I know I've arrived at the place where druids of forest research make offerings to Rot.

Whole watersheds of old-growth western hemlocks and Douglas-firs that grace the Andrews are simply shocking compared to the second- and third-growth evergreens of my home hills. The Decomposition Plot, devoted to studies of nutrient cycling and forest refreshment, lies in one such ancient stand. It's easy to tell when I'm inside the research zone by the yellow, red, and blue tags on wire stems sprouting from the moss. One pink cluster pokes like old trilliums from a mossy mound that once was a tree. A red bunch limns the ground where a onetime log has finally given up the ghost. Metal tags label the cut butt-ends of many logs that lie about higgledy-piggledy, as gravity and the wind might have arranged them had researchers not dropped them first. Bright flags beribbon trees, shrubs, small boles, and limbs, and duct tape shores up the ends of some logs: is someone investigating the degradation rate of duct tape as well as wood fiber? White plastic pipes, buckets, jugs, and other bits lie here and there, each significant to some experiment or other. In early spring, no one is here for me to ask.

Some would see all these artifacts as litter, marring their wilderness experience. You can also see them as inflorescences, like that mysterious white plastic funnel sprouting next to a nodding trillium. Take away the pink ribbon around that hemlock over there, pick up all the aluminum and plastic, and this old-growth forest would still work just like any other. Researchers cut fresh cookies for a starting point, then measure their decay forever after—or as long as they can. But let all the straight cuts rot away and you've got an untidy place going about the important business of trading in the old for the new, an ecosystem definitely in it for the long haul.

For the most part, most of us take the short-term view, most of the time. What gratifies right now, or soon at the latest, is always more compelling than what might satisfy years from now, let alone nourish the generations. When business opts for short-term profits instead of long-term husbandry, both forest and human communities suffer. The short view is what turned most of the Northwest's giant forests into doghair conifer plantations cut on short rotation for pulp. To peer much further down the line requires not only empathy for those who follow, but also faith in the future—even if you won't be there to see it for yourself. Such an ethic underlies all of the long-term studies here on the Andrews, whether concerned with old-growth ecology, hydrology, riparian restoration, forest development and mortality, carbon dynamics, invertebrate diversity, or climate change and its effects.

Meanwhile, here in the Decomp Plot, nuthatches toot in monolithic columns of Douglas-fir; a robin chitters in a clearing. Dappled light falls on forests of the moss called *Hylocomium splendens*, hammocks of shiny twinflower leaves, and fleshy *Lobaria* lichens lying about like tossed-up ocean foam. The path is a maze of Irish byways for voles. Douglas squirrels leave their middens of Douglas-fir cone bracts all about like a prodigal's spent treasures, and round leaves of evergreen violets and wild ginger spatter the path like green coins. If they were gold, I doubt they'd distract the unseen leprechauns who come here to gather the data of decline. Gold doesn't decompose, and this place is all about the documentation of rot. It goes on all around me: something fairly large just fell from a nearby old-growth giant.

Maybe that's the problem with the long view: it speaks of our own inevitable demise. We're not much into self-recycling. Even in death, we take heroic steps to forestall rot by boxing our leavings in expensive, hermetic containers. After all, to anticipate the future—a future without us—is asking quite a lot. But life and regeneration are the name of the game on this mortal plane, every bit as much as corruption. The winter wren's song, after all, is no morbid message. Old vine maples hoop and droop under their epiphytic shawls, but the unfurling leaves of the young ones are the brightest items in the forest (even brighter than the red plastic tags). Every downed and decaying cylinder of cellulose makes yards of nitrogen-rich surface area for hopeful baby hemlocks, lichens, liverworts, and entire empires of moss to take hold on and begin making forest anew.

If we care about what's to come, it makes sense to send delegates to the forests of the present to find out how things truly are, report back, and check in again year after year. The conundrum of the diminishing baseline says that if we have no clear idea of what went before, we are more likely to accept things as we find them, no matter how degraded they may be. Memory is short, the collective memory even shorter. But with baseline in hand, we can appreciate change for what it is. Recognizing loss, we may even act to prevent future loss.

Just as the scientists gather data, any open-eyed observer could go on documenting details without end in such a place: the declination of that row of saplings bent over one deadfall by another; the way that one sword fern catches the sun to suggest a helmet; how the polypore conks launch out from cut ends as soon as they can after their vertical hosts go horizontal, their mycelia reorienting ninety degrees to the zenith. There is no end to particulars as long as the forest goes on and there is someone to record them. The moss grows, the raven barks, the trees go to soil—first hemlocks, then firs, finally cedar. All the while, the decomp team is there, watching how the cookies crumble. Maybe looking to the future is a way of hoping there will still be something to see when we get there. Maybe it's the only way to make sure of it.

The Web

ALISON HAWTHORNE DEMING

(with lines from Claude Lévi-Strauss)

Is it possible there is a certain
kind of beauty as large as the trees
that survive the five-hundred-year fire,
the fifty-year flood, trees we can't
comprehend even standing
beside them with outstretched arms
to gauge their span,
a certain kind of beauty
so strong, so deeply concealed
in relationship—black truffle
to red-backed vole to spotted owl
to Douglas-fir, bats and gnats,
beetles and moss, flying squirrel
and the high-rise of a snag,
each needing and feeding the other—
a conversation so quiet
the human world can vanish into it.
A beauty moves in such a place
like snowmelt sieving through
the fungal mats that underlie and
interlace the giant firs, tunneling
under streams where cutthroat fry
live a meter deep in gravel,
fluming downstream over rocks
that have a hold on place

lasting longer than most nations,
sluicing under deadfall spanners
that rise and float to let floodwaters pass,
a beauty that fills the space of the forest
with music that can erupt as
varied thrush or warbler, calypso
orchid or stream violet, forest
a conversation not an argument,
a beauty gathering such clarity and force
it breaks the mind's fearful hold on its
little moment steeping it *in a more dense*
intelligibility, within which centuries
and distances answer each other
and speak at last with one and the same voice.

Scope

Ten Small Essays

JOHN R. CAMPBELL

1. BLOW DOWN

Can't enter the woods directly. Too dense, too many snags. Cultural clutter. Overcrowded mind. So I sidle in, to where young fir trunks are downed by the dozens, snapped and slung to the ground by wind. At first glance it's chaos, but soon pattern emerges: a cross-slope, mostly northeast orientation. Some perpendicular trunks as well. Thicker trunks are severed higher, thinner trunks broken lower.

Each snag is a sentinel to a fallen self. And each log is host to subtle fungi, some white and amoebic, some black and gnarly, some apricot and globular. Strewn twigs, little arcs, grace the moss and the sword ferns. (O the resilient ferns—underleaves rough with double rows of tiny, tawny spores.)

I move upslope to see: a lattice. Wood in airy layers. Wreckage suspended, like a promise, just inches over the soil.

2. BRIC-A-BRAC

Here and there in the experimental forest: quirky human artifacts. Plastic non sequiturs. Buckets, screens. Pink or orange ribbons. Spray-paint on trees. Tarps spread on the ground in forest groves. Little buildings, gauges on streams. For science, this is the bric-a-brac of inquiry. Though the exact functions of these paraphernalia remain (appealingly) obscure to me, quantification is the general

idea, yes? The oddness of these objects—in the context of the forest—bespeaks the riddle-solving quest.

Seeking patterns amid complexity, science practices anomaly. As do artists. As do poets. And the writers of scrawny essays. Scientists want data. Artists, what do we want? I hesitate to rush toward an answer here. The question requires research.

In the field.

3. WORLDS, REALLY

Hiking the old-growth trail at Lookout Creek, I descend, ascend. The trail, following the contours of Lookout Mountain, wends its way through archetypal forest. The March sun, just past equinox, angles down through true fir, Douglas-fir, western redcedar.

If I step lightly in these woods, it's not because of my mood. The ground beneath me is soft with deep forest debris. In places it's almost buoyant: walking atop a massive log, my feet sink into a spongy pulp. With the next step I'm lifted an inch or two, only to sink again. This, my lilt, my gravity, my comical dance with decay.

I experience the forest not as an expanse so much as an arrangement. It's a vertical landscape, where levels and layers supplant distance as the focus of attention. The great height of the trees among shifting mountain elevations intensifies this effect. Ascending, descending, through various zones of temperature, humidity, and light, the trees serve as organic gauges: what life-forms are possible, what variety and beauty, within given conditions. The lichen-draped, pinkish, sinewy bark on the buttressed trunk of a cedar. The warm brown, gorgeously rough trunk of a Douglas-fir.

Here individual trees, which we typically perceive as figures, are so immense as to become grounds. They are platforms for myriad life-forms: ferns, mosses, lichens. Fishers, chickarees, chorus frogs. Chickadees, woodpeckers, barred and spotted owls. Even (or especially) when dead, the trees are grounds for life. Rotting logs feed beetles and fungi. They also sport tree seedlings and, eventually, saplings. As nurse logs, they provide structure. They lift the seedlings above the deepest shade of the understory. They retain moisture. They generate soil.

The trees craft the conditions for their own thriving. Old-growth trees are worlds, really: they create their own climates, and they eclipse the forest floor.

4. ODE TO RHODODENDRON

Litter of scrolled leaves, leather-brown. Look about: rhododendron trees. Where a gap in the canopy lets down sun, the leaf stems are intensely yellow. Rhododendron leaves. Green-yellow. Aglow. Oblong and elliptic. Drooping, slightly. Arranged in a circular pattern from the end of an elegant branch, the leaves are parasols, are celebratory. But quietly so: O soon the erotic buds, the siren pink flowers.

5. TIME/BEING

From informational signage at the bottom of Forest Road 1506:

> In geologic terms, the landscape of the Cascades is young. The oldest rocks are made up of pyroclastic material (ashflows, mudflows, and breccias) associated with a period of volcanism and uplift twenty to twenty-five million years ago.
>
> About four million years ago, more recent volcanic activity began to shape the landscape we see today. Beginning again with pyroclastic eruptions near present-day Lookout Mountain . . . lava coursed down river valleys. . . .
>
> The landforms of the Andrews Forest may well be an example of "inverse topography." A general uplift of the region, associated with mountain building, helped accelerate the downcutting action of the developing stream system into the older, more erodible, pyroclastic deposits.

Clambering down into another steep creek bed, I'm stopped where a feeder stream has inundated the way. As I negotiate the sodden trail, one deliberate step at a time, I remember: I'm negotiating not only space, but *time*. Moving through time is nothing new, of course. Time, expressed spatially, is usually horizontal, linear. Time moves forward or, in fantasy, back. Yet the slope of this mountainside belies that trope as time expresses itself in the earth. Here that expression is mostly vertical. It has texture and depth.

If my cursory understanding of the local geology is correct, I'm not only descending into a creek cut, I'm entering the original pyroclastic flows. I'm stepping, unsteadily, down (not back) toward twenty-five million years ago—a sneeze in geologic time, but still . . .

Having lived in Utah, I'm used to traveling canyons, moving through naked geological time. But here, the densely vegetated mountains secret their mineral faces. I catch a glimpse, deep in the creek bed, obscured by roots and ferns and delineated by water, of mineral frankness. Grayness. Blank expression.

On a geological scale, time rumbles upward: mountain lifting. Time etches downward: erosion. On a biological scale, time soars upward, via these trees, some of which are two hundred feet tall and over five hundred years old. Inside the boles, past forms of the trees have been preserved in near entirety. I know this only because an Italian artist, Guisseppe Penone, once chiseled into a vast wooden beam, working his way down, in three dimensions, to the contours defined by a specific ring. What he revealed was an exact sapling, the tree as it was, and is.

Moving through these layers and scales, I experience time as physical, specific. Time loses its dreadful abstractness. No longer an external force, time is the Earth itself.

6. VARIED THRUSHES

Oregon robins, they used to be called. At my approach they start from the forest floor, where they've been foraging. Their wing beats are audible as they hurry to nearby branches, where they're enticingly obscure to me. I know their bold markings: slate gray and orange, the male with a black mask and breast-band. I spy one in an old Doug-fir and glass him as best I can. As I bring him into focus, I'm thrilled by those fierce markings. To me, they express survival, a ubiquitous presence in coniferous woods. The scattered birds want to regroup. They begin uttering their contact calls. One note: subtle, metallic, yet sweet.

7. ABANDONED CAMPGROUND

I learned long ago to visit abandoned campgrounds. Closed for the season. Half-burnt wood in the fire pits. Picnic tables stacked, dumpsters laid on their sides. Forest litter beginning to obscure the asphalt. To be where people are not. This is, often, my ambition when I'm in the woods. It's not the same as being alone, nor is it lonely. There are presences enough in abandoned places.

My craving for solitude is comically dire. It's a neoromantic flaw. Epicurus wrote, "We know not death, for when we are here, death is not. And when death is here, we are not." My craving is kind of like that, though not so severe: when I am here, people are not. And when people are here, I am not. Sometimes I like to just miss them.

Delta Campground is gated. I've parked the car, careful not to block the gate—I'm a courteous sort of antisocial—when an old Volvo pulls in behind me. A man and a woman, young, pleasantly disheveled, gape at the locked gate. The driver rolls down the window, says, "Do you know of anywhere we could camp around here? This is the third place we've tried. They're all closed."

"Well, I don't think they open these campgrounds until Memorial Day," I offer.

"Don't they know it's Spring Break?" the driver asks.

I just smile and direct them to the town of McKenzie Bridge, up the road. "You could ask around there."

In the passenger's seat, the woman is pointing at some spot on a Forest Service map. The driver nods to her and begins to roll up his window. Remembers his manners. Says, "Thank you, sir," and drives off.

The "sir" part kind of grates on me, I guess. One thirty-second encounter with humanity, and I've already forgotten the lesson of time.

In the campground, the Doug-firs are plenty big, two hundred feet tall, some 650 years old. They skirt some McKenzie River side channels. Lime-green algae wavers in the current. Beneath a fallen fir that spans the water, a pair of mergansers moves through.

8. BLUE RIVER RIDGE, IN SUN

Bees hum, seeking manzanita. Contrails drift. Black spiders shuttle in the scree. Sometimes place just scratches out words. One word at a time.

9. ROOM IN THE FOREST, IN RAIN

It's the second day of solid rain—and here that adjective is not merely pat. From my room at the Andrews, I'm gazing out through two layers of rain. The first is composed of thick, sporadic drops from the eaves, and the second is a fine and steady screen. And this against a tapestry of green. There's an attempt at a lawn, tufts of grass mixed with chartreuse moss, where a flock of robins is foraging. Then, a dense layer of young firs, darkness—and old stumps—beneath their hems. Pillars and snags just beyond, the old growth rimming Lookout Creek. On the black ridge above the old tops, clouds climb and rend all day.

Classic Pacific Northwest. Intimations of Chinese art as well. This scene is embedded culturally via images, the province of artists and poets. And yet: in the experimental forest and beyond, the exploitation of "natural resources" is as graphic as a clear-cut slope, or as subtle as silting streams. Once the research at the Andrews Forest served the naked exploitation of the woods. But over the years, research at the Andrews has moved toward sustainability.

Informed by such research, we manage for entire ecosystems now rather than individual species. In some very important ways, we are capable of change.

Still, a certain status quo persists: nature as instrumental. (Instrumental: describing a noun case that indicates something is used for a purpose or is the means by which something is done.) Commerce insists on it. Cultural stasis maintains it. But human-induced environmental change has now presented itself on an unprecedented scale. What we're witnessing is not "the end of nature"—because that trope suggests an artificial separation of nature from human culture—so much as, to quote Jane Lubchenco, "Humanity . . . as a major force of nature." Given this reality, what new cultural urges might serve? Global warming, habitat depletion, species extinction, overpopulation: these crises tax our moral capacity, which is to say our imagination.

Imagination. Has the term become quaint? Nevertheless, it's fundamental. Empathy is a function of imagination. It is also the origin of ethics. How well do we imagine the lives of our human others, including future generations? How well do we imagine the lives of our animal and vegetable others? The answers to these questions will determine our ethical choices.

Perhaps the greatest challenge to contemporary environmental imagination is our ability to conceive larger temporal and global scales. Addressing global warming, for example, demands such an ability. In this respect, art can learn from science. The Andrews Experimental Forest is an integral member of the Long-Term Ecological Research Network: "a collaborative effort involving more than 1800 scientists and students investigating ecological processes over long temporal and broad spatial scales." Research at the Andrews involves, for example, rates of forest decomposition over a two-hundred-year span.

Artists can learn from science to extrapolate from the specific to broader scales. By broader scales, I mean not grand assertions, nor metaphysical slants. I mean an imaginative entry into the physical vastness of time and space.

Environmental science benefits from data-gathering technology: global positioning systems, satellite imaging and mapping, isotope tracing, remote sensors, and so on. Scientists get data in the field and feed it into computer programs that simulate changes, usually graphically, over time and space. In other words, science employs present-day data in order to model the past and, for predictive purposes, the future. This is one version of imagination, yes?

Science also offers insight as to how the present, living planet might contain all possible pasts and futures, allowing for both change and constancy. Oliver Morton, in *Eating the Sun: How Plants Power the Planet*, explains:

> The earth is not brute matter. . . . In flames and hurricanes, whatever might happen next can depend on only what is happening now. For the [living] planet as a whole, whatever happens next depends on everything that happened in the past, because the record of that past, and the means to reproduce its processes, are locked away in molecules that may be accessed and used at any time. Extinctions may remove species and shapes and behaviors, but they do little if anything to biochemical possibilities. For as long as they have a use, the planet will never forget the workings of [photosynthesis]. Or the fixation of nitrogen, or the trick of being a tree, the sense of when to relax a leaf's stomata. Unlike a simple flame, the planet can be tomorrow what is not today, and the day after change yet again, while all the time remaining itself.

Oh: perhaps nature *is* instrumental, on a scale so vast we are uncomprehending. Though we may bend it to our will, and radically diminish it, ultimately it meets *its own* imperatives, not ours. It does so not consciously so much as inevitably. So to shift our attention from human demands to Earth's requirements is a crucial cultural task, one the artist might well undertake.

How might I begin? Today, rain or no, I walk in the woods. I let the woods direct me. Instinct and science convince me: in this specific present is all the world I require.

10. GUEST

After days of wandering through some of my favorite haunts—old-growth trails, abandoned campgrounds, obscure Forest Service roads—I've arrived at a place of calm uncertainty. The woods here are stunningly complex, but they don't overwhelm me so much as absorb me, easily, into their textures. So to be *absorbed,* which to me was for too many years a matter of thought and identity, now becomes an honor. I am received.

The host that receives me is no person, of course, and no personified god. It's a host in this sense only: the forest actualizes multitudes and allows me to walk among them.

Let me mention the two finest pleasures I've had in my time at the Andrews. First, an absence. Second, an anointment.

* * *

Toward the end of the old-growth trail, I encounter sporadic patches of snow. Where snow has bridged a stream: animal tracks, indistinct. Where the imprints have thinned the snow, the melt is slightly accelerated, I suppose. The tracks are deepened and blurred. Bobcat, maybe, by the size of them. Are those four clawless toes? (Cats retract their claws while walking.) Soft edges make it hard to tell. Wait: is that a rounded heel-mark? If so, it could be red fox. Or, for all I know, both animals could have come this way. Suddenly, pleasantly, identification is no longer imperative.

And look: one of the tracks has become a negative shape, revealing the earth beneath. Within the shape of the track I glimpse the edge of a fern frond, and moss, and needles. I marvel at this emblem, where animal, plant, and precipitation are merged. Where the animal's absence is not only a presence, but also a potent, momentary image.

* * *

Later, strolling near the Andrews headquarters, I admire the Pacific silver firs. The bark is patchy, from gray to gray-white, and knobby, adorned with tiny blisters. These little oblong bulges: I remember reading of "resin vesicles" in the bark of young firs. With my thumbnail I slice a blister away, and yes—it's filled with a dram of liquid resin, which spills onto my hand.

This is not sap, oppressively sticky. It's more a tacky oil. The fragrance is beyond description. It edges memory: I'm in my wife's art studio, talking with her while she cleans her brushes with turpentine. I'm tasting a certain liqueur. The correspondences aren't exact, but associative. Christmas trees, of course. But something more: soap? Or grape on the very margins of the scent?

With a washing motion, I rub the resin into my palms. I want to feel the air on my skin. I want the resin to mix with my own body oils. I bring my open hands to my face.

Inhale.

GROUND WORK

Natural History of the Andrews Forest Landscape

The twenty-five-square-mile Andrews Forest is nestled on the western flank of the Cascade Range and occupies the entire watershed of Lookout Creek. Elevation ranges from thirteen hundred feet, where Lookout Creek flows into Blue River, up to the summit of Carpenter Mountain, at fifty-three hundred feet. The underlying bedrock is composed of lava flows and mud-flows cut by dikes during several episodes of volcanism over tens of millions of years culminating about 3.5 million years ago. Ice Age glaciers carved the headwaters of the valley, and the pervasive effects of landslides and stream erosion have sculpted the present mountain terrain. These landforms are the physical stage on which terrestrial and stream ecosystems perform. Geomorphic (land-forming) processes, such as floods and landslides, are periodic agents of disturbance.

Winter storms from the north Pacific encounter the west slope of the Cascade Range and annually dump an average of nearly one hundred inches of precipitation on the Lookout Creek valley, falling mainly as rain at low elevation and snow up high during the winter wet season. Large floods periodically occur when heavy, warm rains fall on freshly accumulated snow. Such floods can trigger small landslides, tear up riparian vegetation, cut new channels, and transport massive logs. But the same streams in summer have tranquil, low streamflow, limiting the extent of surface-water aquatic habitat.

A small number of evergreen conifer species dominate the tree flora: Douglas-fir, western hemlock, western redcedar, true firs at higher elevation, and the enigmatic Pacific yew, whose thin bark is the source of the cancer drug taxol. The evergreen character makes possible year-round photosynthesis, although at reduced rates in the short, cloudy days of winter. Large tree boles serve as water reservoirs during the dry summer. Thick bark and limbs high above the forest floor render the trees rather resistant to wildfires, which occur when the unusual confluence of lightning ignitions, exceptionally dry forest fuels, and hot, dry east winds push wildfire across the mountain landscape. Both tree physiology and infrequency of intense fires make possible the great age of trees; many of the largest Douglas-fir date from a regionally extensive episode of fire approximately five hundred years ago.

But the forest is much more than massive, ancient trees. Wildfire in the mid-1800s created extensive patches of what now make up a "mature" age class of forest (for a tree, that's eighty to two hundred years old). Clear-cut logging in the 1950s and 1960s resulted in plantations of young Douglas-fir trees that now occupy about a quarter of the Andrews Forest landscape. Wet and dry meadows dot the highest slopes. The resulting landscape is a mosaic composed of vegetation patches of varied origins, plant communities, and dates of initiation. Herbs, shrubs, and small deciduous trees, such as vine maple and dogwood, are common throughout the landscape, ranging in expression from subtle understory communities to dense, impenetrable patches. In these diverse vegetation conditions, thousands of species of invertebrates find a wide variety of habitats, from deep within the soil to high in the canopy. Hundreds of lichen and moss species drape the limbs of trees and cling to the bark. Vertebrate species are surprisingly limited, though cougar, bear, bobcat, and over 170 bird species reside in or visit the Andrews over the course of the year. This assemblage of organisms produces a relatively quiet biophonic soundscape; animal sounds provide only limited choral complement to the wind soughing through the evergreens and the reverberation of streams cascading over boulders.

Although they occupy only one percent or so of the landscape, streams fulfill many critical functions in the environment. While streams are intimately interconnected with the surrounding forest, they are also a distinctive ecosystem with obligate species, such as fish and a component of the amphibian community. Other vertebrate and invertebrate species are "shared," because their life histories carry them between streams and forest habitats.

Change is the ubiquitous feature of a steep, wet, vibrant forest landscape such as Andrews Forest. One can sense change ongoing in the tattered forms of the giant trees, the seasonal behaviors of plants and animals, shattered fragments of the fallen giants with their upturned root wads, the coming and going of mushrooms, and the ever-shifting patterns of light, sound, and fragrance.

Threads

VICKI GRAHAM

From the larger poem *Debris*

Light gatherer, shade tolerant climax tree,
the western hemlock spreads needles
flat to the canopy.
Underneath, a coralroot's orange stalk
draws light-manufactured sugars
from the hemlock, opens
to a spray of red and purple striped orchids.

* * *

Creek music: thunk thunk thunk
deeper than the heartbeat of stones,
a bass line thrums
under the splash and cymbal crash
of water splintered on rock.

* * *

Firs weave *now* into seed and cone,
root and hypha, needle and branch,
layer *then* thick as duff
on the forest floor.

* * *

Seamless, water opens, slides
over rock, closes again.
The flung drops scatter white,
catch sun, foam,
vanish downstream.

* * *

Moss hangs like mist, drops scrims
from maple and yew,
hemlock and fir, filters sun,
filters time. Silent and old
as the trees they grow on,
moss tapestries began
as single spores caught
in a sapling's leaf-scar.

* * *

Roots grow deep, not open
in light and air, but lapped, cramped, crowded,
jammed and crossed, rigid, ingrown;
arthritic fingers clutch earth, seek water, cradle stones.

* * *

Gravel bar: without the bead lily's hexagram
or the fir cone's logarithmic spiral—
without the sonnet's symmetry
or the haiku's clipped syllabics,
winter floods pack cobble in patterns.

* * *

Under the paper thin skin,
the yew's inner bark burns red—
black coals broken open, glowing.

* * *

Vine maple winds through the forest,
turns hands flat to the sun.
Light pools in the palm the way water pools
in the lungwort's lobes.

Interview with a Watershed

ROBIN WALL KIMMERER

October 28, 2004, 1600 hours. Data points come up on a computer screen at the Forest Sciences Lab in Corvallis (0.162 14.3 12.0 0.123 9.34) fed from a T2 line running down the valley of the McKenzie River from a telemetry station at the H. J. Andrews Experimental Forest. The numbers arrive at the telemetry terminal as a radio signal transmitted from a small box of wires out in the woods, where a chipmunk sits on the cover absorbing the modicum of heat from within. Wires run from the box to sensors that rest in the riffles of a burbling brook.

On the same day, at the same time, I am sitting beside that brook resting on a mossy stone. The radio transmitters are silent and all I hear is water, trickling down, ledge to ledge over mossy boulders and sluicing under logs. Down next to Lookout Creek, the sound is a white-noise roar; but up here in the headwater stream the many voices of water are heard, low gurgles under rocky ledges, high notes of small rills, and the bell tones ringing from deep green pools.

The elders used to say that you could learn a lot from listening to water. It will tell you what you need to know, what has happened before and what is on the way. My friend Frank Lake, a Karuk from the mountains to the south of here, tells me that his people still make a circuit to all the springs and streams in their homelands, to check on the health of the land. They taste the water, watch its flow, and see how thick the plants grow. They clear any sediment from the springs and look for the Pacific giant salamander, a sign of the waters' well-being. At each pool, they offer prayers of thanksgiving for the waters and in hopes that they will continue to run. Long ago, and to the present day, our people did not turn to sacred texts for understanding. We understood back then that wisdom lived in the land.

Set in a cleft between two slopes is the gauging station, a two-story dollhouse painted red with a moss-covered roof. We stand on its miniature porch and John describes the weir below, a broad concrete V that lies beneath the stream. The water flows right from the mossy stream, across the weir, into a rocky pool, and then again into its native streambed of rocks and fallen logs on its way downhill to join with Lookout Creek. John Moreau, a technician at the Andrews Forest, unlocks the door and we step inside. He's a strong and wiry man with salt-and-pepper hair and a youthful twinkle. He's been collecting data at the Andrews for twenty-eight years now, and the peace of the place has rubbed off on him. He's been part of the team from the days of clipboards and paper-strip charts to radio telemetry. Summer and winter, he installs and maintains instruments and collects data from a network of sensors all over this watershed. In summer, it takes just a day or so a week to collect all the values; but in winter on the snowcat, it can take twelve hours just to retrieve a single set of samples. He unsnaps the cover on the water sampler, which stands in the middle of the room. Water samples are automatically sucked in from the stream, through tubing that runs up through the floor to an intake port in the pool outside. Mounted nearby, a telemetry box with its nest of wires, radios stream data from the sensors: flow rate and volume from the weir, temperature, and oxygen levels . . . data already on their way to Corvallis. The second floor of the little house is fitted with a well tank and a system of weights and levels to accurately read the level of the stream.

We climb a slippery trail winding up the steep slope above the weir to where a tower made of metal struts juts up through the canopy more than a hundred feet above us. At intervals along the tower are more sensors, pyranometers to measure light, thermometers, anemometers to measure wind, psychrometers for humidity, and sensors to measure atmospheric gases at different heights in the forest canopy.

Day after day, raw data streams from these sensors in a flow of electrons, representing the flow of this water. It used to be that people harvested trees from these slopes; today they harvest data. Input to the forest is measured as precipitation, output from the forest as streamflow at the weir, creating a hydrologic balance sheet. But the accounting doesn't add up: there is water unaccounted for, and now the researchers are looking for it. White PVC pipes stand up from the ground in arrays along the streambed, hyporheic wells that allow researchers to measure the invisible flow beneath the surface. Wires run out of the ground, connecting to soil-moisture meters. John describes to me the next step in inventorying the movements and interactions of the forest's water. This season he will be installing sap-flow meters on the trees, tiny thermocouples designed to detect the flow rate of sap as it rises up the tree trunks on its way to the atmosphere. One year they rigged some mossy branches high in the canopy with strain gauges to determine the weight of water captured by the mosses.

John tells me that when he began here in 1976, loggers were cutting old-growth forests at a furious pace. Log trucks full of ancient trees were barreling down the valley at a rate of one per minute. The Forest Service, the College of Forestry, and many folks at the Andrews were embedded in a culture of board-foot forestry. A few had the wisdom to challenge this thinking, leaders at the Andrews among them. At a time when scientific forestry viewed old-growth forests as decrepit liabilities, scientists at the Andrews set out to understand the influence of the presence of such forests and of their absence as they fell to the saw. This did not make the scientists popular in the valley.

This weir sits at the bottom of one of three small, paired watersheds. It measures all the water that drains from this forest through the trickling stream. One of the three watersheds is intact, an old-growth stand of massive cedars and firs. Another was partly cut, in patches, and this one where we walk was clear-cut. The logging company brought in a Swiss team with a new technology—new back in 1962—for skyline logging, to remove the trees without the damage of roads on a slope so steep you can hardly stand up on it. I ask John why they cut on such steep slopes. He looks at me quizzically and says, "That's where the trees were." Every tree was cut and hauled away, leaving a bare slope behind. They planted Doug-firs on the slopes, and built a weir and gaging station on the stream. Day by day it sent out data on water flow and chemistry, data that told the story of a landscape hemorrhaging nutrients and filling the stream pools with sediment as the soil washed away, down to Lookout Creek, where it silted up the fish spawning beds. Sensors recorded the increased temperature of the streamflow, warming in the absence of the shading canopy, too warm for trout and salmon. Meanwhile, over at watershed 2, still covered with old growth, the stream ran cold and clear and pure.

Water is a storyteller, and listening to that story helped to write a new one, in which old growth has a role. It is a story nearly too late in being heard, but now there is a chance. These studies have been pivotal in changing our thinking about forest management, in understanding the connections between what we sow in the short term and what we reap over time. The opportunity lies in listening to the land for stories that are simultaneously material and spiritual. It is a hopeful sign that people return to the words of the elders and again look to the land for knowledge. Our people say that long ago we could all speak the same language—the trees, the birds, the wolves, and the water—but that we have long since forgotten. Human capacity has been so reduced that we understand only our own tongue. I like to think that, in the right hands, scientific research is a conversation, an interview of sorts between two parties that don't speak the same language.

Lewis Thomas has said that humans have four kinds of language. The first he says is chitchat, the patina of words we use to coat social interaction; the second is conversation, real talk where information and ideas flow with energy between two minds. The third type of human language is mathematics, a higher-order code that transcends dialect and ambiguous interpretation. Mathematics is the language we use to interview the land. We cannot readily converse with the forest about what makes cedars grow slower when the temperatures rise. But we can ask questions. We can slide a sap-flow meter under the bark and measure the rate of water uptake at the same time that a digital meter inquires after the amount of water in the soil. Tree bands tight around the girth of the cedars yield readings of changes in diameter that indicate growth. We can read the temperatures from the meteorology tower and chart it all out, looking for the patterns that will tell us what cedar needs to flourish—and what might happen if the temperatures rise. The sensors and the weir, I think of as a microphone amplifying the voice of the water and translating it into numbers so that we can try to understand. But there is danger in thinking that we do understand. We cannot say to the forest, "Did you suffer terribly when the trees were all gone?" But we can measure the hemorrhage of nitrate washed away. We might want to ask about forgiveness, but instead we measure the increasing clarity and oxygen of the stream, and hope that it will suffice. Data alone do not bring understanding. You can collect data in a day, information over a year, knowledge over a decade, but wisdom takes a lifetime. Or more.

The digitized flow of data packets from the stream back to the university is quick and efficient and allows a massive pile of numbers to accumulate. And the scientist can do the work without ever getting wet. I'm not sure that's a good thing. There is, of course, the problem of relying on batteries and wire to accurately sense the world. John is always screening the data flow for what he calls "wowies," anomalous readings in the data record that make you say "wow!" and go check to see if a weasel has burrowed beneath your temperature sensor. Isn't something important lost by having the data stream back to the terminal untouched by human hands? A column of data doesn't leave much room for surprise. The sensors and their numbers can answer only the questions that we ask, and in the limited way we ask them. The data that the sensors collect might not be all that the land has to say. In conducting an interview, the good reporter asks questions and records the answer. But the real value of the interview comes when she reads her subject's body language, when she looks into the eyes of the subject and sees a truth that is different from the words. You can't see such things if your only way of knowing is data. Can you really understand a place without kneeling in the humus or standing quietly to watch the alder leaves drift down the stream? Being there, doing the fieldwork, is for me a way of becoming intimate with the place, really listening to the land. It makes for better science, because the land will suggest new questions. It makes for better scientists, too, because the land is more than data and we are more than data analysts. Most of us scientists were drawn to our work not by the love of data but by love of the land.

The stream is not yet full to its banks; the flow through the notch of the weir is only a foot wide. At the height of the winter rains, it can be more than five. John shows me the sampler for water chemistry and the bottles that will go back to the lab. He handles them with care, knowing what they contain. "This is the very best time of year for water sampling," he says, "after the second rain of the season." The first rain soaks into the summer-dry soil and is held there in a sponge of humus, rain filling all the pores. But the second rain comes and flushes out some of the soil, carrying it to the creek and to the sampler. It carries messages about the soil; the dissolved nutrients that resided there all the rainless summer are now mobilized in water.

* * *

The number 0.162 flashes by on the screen, only one of hundreds of data points. It seems like a lot, and the researchers employ a team of data managers just to archive it and access it for analysis. But each data point is much bigger than a point; it is a line, a thread that, pulled, goes deep into the forest. Each number floating on the screen is one word of a story: 0.162 ppm nitrate in stream no. 124 is only shorthand for the nitrogen that was pulled from the air by the blue-green algae within the epiphytic lichen *Lobaria oregana* that grew on the moss that cushioned the eggs of the last spotted owl in the valley.

And today's dip in the amount of soluble phosphate, classed as "noise" in the data, just a random variation, is not noise at all. It is the birth of a patch of coralroot orchids, whose network of mycorrhizal roots is scavenging phosphorous from a decaying log. These are the stories told in the water. Ravens scavenging a carcass, a thousand-year-old fir year tree falling, all are held in a data point of nitrogen concentration on October 26. If the voice of every drop of water—every alder drip and maple drop—is altered by its relationships, imagine the stories that a stream has to tell. The data from the Andrews are translations of the stories told by water: things that have happened, things that are coming. Just as our elders suggested, we must listen to water.

The watershed clear-cut thirty years ago is now a three-layered forest. There are massive stumps of western redcedar and fir three feet across, reminders of those who are gone. Overhead is a thicket of red alder, their light bark just beginning to be masked by mosses. The third stratum is young Douglas-fir. Foresters used to think that alder was a weed and did all they could to suppress it. Now, we know how important it is in rebuilding soil, replenishing nitrogen so that the forest can recover.

When we compare the clear-cut watershed to the old growth, the stream tells a very different story. The young, recovering stand is adding nitrogen to the system, by fixing atmospheric nitrogen into leaf and root for the future. Now that the trees are regrowing, no nitrogen is present in the water samples at the weir; it is all being used within the watershed. Over at the gauging station on the stretch of creek that drains the old-growth stand, the water is cold and clear. The undisturbed old growth retains its nutrients, holds on to its deep soils, and slowly recycles its nutrients. By means of all the meters and sensors, watershed chemists have noticed, however, that in some of the oldest stands, there is a slow trickle of nitrogen from the forest. They hypothesize that old forests may accumulate more nitrogen than they can use, and so it is released into the water, going somewhere else where it is needed.

Some forests can become nitrogen saturated. Likewise, we scientists can become saturated with the rivers of data we generate. And what do these data bring us? A chronicle of the land, a witnessing of the world, understanding and wonder, a way to predict our impact on the land. These are good things. But do they bring us any closer to saving what we love? I want a flow of data streamed into some monitoring center that measures a change of heart. I want us to see clearly the jagged peaks of rising greed and their correlation with loss. Shouldn't we make models that predict the conditions under which destruction occurs so that an alarm will sound, shrilly warning us back from the brink? Couldn't the engineers give us special anemometers to detect dangerous shifts in political winds, atmospheric recorders that analyze the sighs of loneliness we feel when the only living beings we encounter are ourselves? The experiments we need to do are about how we can live and not hurt land. How we can heal the wounds that we inflict. For those experiments, I would sit with eyes glued to the terminal, watching for cultural change, in order to chart a rising tide of ecological compassion.

The data we currently collect are valuable and represent a vital piece of our story. But I don't think it is by information alone that we will be saved. My students, once they are filled up with new ecological knowledge and have developed an awareness of our situation, always say, "We have to tell people what's happening in the world. If they only knew what they were doing, they would stop." But, it's not true. We are all saturated with data. We do know what we are doing. And yet we continue, racing headlong toward our greatest loss, wringing our hands all the way. And bleeding from a self-inflicted wound.

It's a hopeful thing when scientists look to the land for knowledge, when they try to translate into mathematics the stories that water can tell. But it is not only science that we need if we are to understand. Lewis Thomas identified a fourth and highest form of language. That language is poetry. The data may change our minds, but we need poetry to change our hearts.

Rich though they are, conversation, mathematics, and poetry are but human languages. And I think there is another language, the forgotten language of the land. Its alphabet is the elements themselves, carbon, hydrogen, oxygen, nitrogen. The words of this language are living beings, and its syntax is connection. There is a flow of information, a network of relationship conveyed in the rising sap of cedars, in tree roots grafted to fungi, and fungi to orchids, orchids to bees, bees to bats, bats to owls, owls to bones, and bones to the soil of cedars. This is the language we have yet to learn and the stories we must hear, stories that are simultaneously material and spiritual. The archive of this language, the sacred text, is the land itself. In the woods, there is a constant stream of data, lessons on how we might live, stories of reciprocity, stories of connection. Species far older than our own show us daily how to live. We need to listen to the land, not merely for data, but for wisdom.

One-Day Field Count

MICHAEL G. SMITH

Seed dropped in the understory tangle, a giant chinquapin fruit burr morphs into land urchin

Snowbrush blazons the plenty of light and nitrogen in the cut and blackened forest

Chinese red, fist-size hunk of Douglas-fir heartwood dribbles itself back into green

Entertaining himself on this day's gravel bar, a boy picks up a chunk of crystallized volcanic conglomerate, asks why a caramel ball chockablocked with butterscotch chips is here, then lobs it to hear the plunk! alongside the walla walla of an eroding mountain's mercurial waters

Thistles springing up on the grassy roadside confirm aliens have gained a roothold

Bird Station no. 204 Temperature and Light Gauge and its relatives are our quiet way of reminding you 24–7 we know you visit these old woods when we aren't watching and want to hear your lovely song, so please show yourselves and we will know when the sunlight and warmth are just right, and will leave the trees alone

Second growth snag with painted blue "L" sings leave me, love me, don't level me

A cedar straddling the border between a blue state and a red state splits its vote

Orange jelly fungi reposed on a nurse log's moist kerf proves that in old growth truth outlives fiction

And the red alder seedlings that have sprouted from the log-becoming-soil carry the infinite Earth-narrative forward

Specimens Collected at the Clear-Cut

ALISON HAWTHORNE DEMING

1. Wild currant twig flowering with cluster of rosy micro-goblets.

2. Wild iris, its three landing platforms, purple bleeding to white then yellow in the honey hollows, purple veins showing the direction to the sweet spot.

3. Dogwood? Not what I know from the northeast woods, the white four-petaled blossom marked with four rusty holes that make its shape a mnemonic for Christ hanging on the cross. This one, six-petaled, larger, whiter, domed seedhouse in the center, no holes on the edges, shameless heathen of the northwest forest that flaunts its status as exhibitionist for today.

4. Empty tortilla chip bag.

5. Empty Rolling Rock can. Empty Mountain Dew bottle. Empty shotgun shell. Beer bottle busted by shotgun shell, blasted bull's-eye hanging on alder sapling.

6. One large bruise four inches below right knee inflicted by old-growth stump of western redcedar, ascent attempted though the relic was taller and wider than me, debris field skirting a meter high at its base, wet and punky; nonetheless, I made my try, eyes on a block of sodden wood, reddened by rain, as fragrant as a cedar closet here in the open air, the block of my interest wormed through (pecked through?) with tunnels the diameter of a pencil. How many decades, how many centuries,

of damage and invasion the tree had survived! But the stump felled me, left me with its stake on my claim and jubilation to see that nothing was mine of this ruin, mine only the lesson that the forest has one rule: start over making use of what remains.

7. One hunk of Doug-fir gray as driftwood, length of my forearm, width of my hand, depth of my wrist's width, wood grain deformed into swirls, eddies, backflows, and cresting waves, a measure of time, disturbances that interrupted linear growth to make it liquid as streamflow.

8. Lettuce lung (*Lobaria pulmonaria*), leaf lichen, upper-side dull green, turns bright green in rain, lobed, ridged surface with powdery warts, under-side tan and hairy with bald spots, texture like alligator skin, sample attached to twig falls at my feet on trail to Lookout Creek. Day five, resampling the site, t.i.d.

9. Four metaphors for the forest. Plantation trees: herringbone tweed. Old-growth trees: medieval brocade. Clear-cut: the broken loom. Clear-cut five years later: patches on the torn knees of jeans.

10. Skat. Pellets the size of Atomic Fireballs, hot candy I loved as a child. This, more oval. Less round. Not red. But brown. Specimen dropped by a Roosevelt elk savoring the clear-cut's menu of mixed baby greens. One pellet broken open reveals golden particles. Light that traveled from sun to grass to gut to ground to mind. Forest time makes everything round, everything broken, a story of the whole.

Forest Duff

A Poetic Sampling

KRISTIN BERGER

boulderfield lookout suppression deadwood blowdown ogive seedling slow-growth rot ouzel pit sunshaft trunk angle emit absorb creak moan pitch sap rise fall mountainpass snow-zone bridge nutrient control-burn sawtooth by-way leaf-chatter gray squirrel dusk whorl basalt burl raven succession burn seed ridge carpenter beetle needle scree ruffed grouse bone-yard clinker pull-out vine burrow ignite cloud-birth valley bore river joint bone-chip fuel success char-pit exposure boundary tangle wash-out fog-bank expiration cold-front climate air-shed reserve impact zone glaciated descend earthstar lungwort ensatina millipede licorice kinglet clubmoss tree tip-pit chanterelle shelf spotted owl first-snow needle-choked sphagnum carpet arboretum seep fan cavity lodge niche riverteeth nitrogen-fixed nurse-log bole shrew snag camouflage bushwhack indian pipe earthtongue release spotted skunk waddle beargrass red-backed vole snowbridges chickadees prince's pine pod violet osprey deer mouse galvanized nocturnal cotton lily basketgrass coral club lobster chipmunk snowshoe hare broadsword turkeytail filament old-man's beard big leaf maple flying cloak Douglas-fir clear story western redcedar earthball intensive lichen nest rapid fog punchbowl cliff corridor juvenile shrapnel canopy drip convergence mature scrub log-jam compost reservoir lateral moraine glacial field station gauge test watershed measure flicker map subalpine rain rough-skinned newt temperate roadkill slide river-bed guardrail milepost yew hemlock flag spruce alder waver white oak ash egg sac rhododendron scour oregon grape riffle bark fiber knoll cradle cut-bank point-bar blind-creek braid channel thalweg filter dipper divergence pebble turtle-back eddy agate ramp flood grade turbulence viscosity flow willow surge siphon mudbank capacity artery fork vascular water-logged respiration tread variant traverse locate molder whet submerge hatch fry larval velocity trip molt upwelling spine migrate return scotch broom false brome columnar cougar savannah madrone bracken salal slash recover plot prey fungi diurnal areal ice-heave plantation cone flush flock grove hibernate cover tremor cache tunnel forage subsist substrata terminus saddle ravine lapse season forge ecotone riprap choke silt updraft flume debris mound dormant stem decay sprout ermine stump pica horizon storage dry-ki gore patch gradient hiatus protection clear-cut year-lings fertile elevation parcel topographical duff hunger monitor predator slope scale blind litter

Pacific Dogwood

(Cornus nuttallii)

JERRY MARTIEN

Ghosts of winter light
pale lanterns floating
in the dark green air
whorls of four or five or six
petals made up to look like
magnolias in May not
actually even flowers
but involucres or bracts
the real flower a small
offering at the center
of a white jade bowl
a cluster of tiny
dark inflorescences
which by summer will be
the red berries favored
by band-tailed pigeons
almost as much as by
naturalists the flower
first correctly ID'd
by Thomas Nuttall
on the Wyatt expedition
to the West 1834
later named after him
by his friend Audubon
who put both flower and berry

in *Birds of America*
with the pigeon
according to Peattie's
History of Western Trees
and not one of those
eminent naturalists
could refrain from
poetic excess
in describing this
bright unveiling in the
darkness of the western
forest any more than I
can resist the desire to
name and know every
detail of that white
breast and its flower.

Riparian

SANDRA ALCOSSER

If we wash our legs with frozen water
Watch it rill down hairy flesh—oh the power
Of the body to refresh—to lie down at night
Wake again among harebells and bees, lichen
Speckled boulders, mists of sweet white
Goatsbeard—if we cock our pollen hats
Like Leonardo da Vinci and sketch
Riffles come to nurse the thirsty
Rubble, we can lean back, sieve
Our tea among secretive
Rocks—soak away the meanness
Of a year's duplicity—no one can reach
Us here—no human noise—
A river will gentle the cruelest voice

GROUND WORK

Old Growth

All the dominant conifers of the Pacific Northwest rain forest—western redcedar, western hemlock, and Douglas-fir—grow together in this grove. The tallest trees reach three hundred feet above the forest floor, anchored by root systems that extend only six feet into the soil below. Some of the biggest trees' trunk-diameters exceed the span of your outstretched arms. Annual-ring counts reveal the grove to be about five hundred years old. Centuries of vegetative growth, species interactions, and storm damage have created a richly complex forest, with understory shrubs and herbs, shade-tolerant saplings and midstory trees, a profusion of moss and lichen, and a wild array of animals living in, on, and among the plants. Light shafts and sun flecks move silently through the forest as birdsong, stream percussion, and the soughing of the wind in the canopy high overhead fill the soundscape.

We now know this as "old growth," or more reverentially as "ancient forest," but in the 1930s, in an assessment of forest resources of the region, it was mapped as "large saw timber." All wood was seen through the lens of board feet. In the parlance of loggers, old-growth forests themselves were "decadent" and "overmature"—they'd been left standing past their prime. What the nation needed was good, easy-to-mill seventy-year-old timber. "Get the cut out" was the mantra. Produce two-by-fours; build houses. Much of that old-growth cut is now out of sight, converted decades ago into studs for the walls of West Coast homes.

* * *

Public perception of western forests has since changed dramatically, owing in part to scientific research that took place at the Andrews Forest. By the 1990s, public opinion of old-growth forests had shifted to the point that the ecological values of intact forests were increasingly evident, and many regarded ancient forests with reverence. The desire to preserve old-growth forests—along with their concomitant, and sometimes endangered, species such as the northern spotted owl—has led to a profound shift in federal forest policy, from the timber era to what might be called the biodiversity era.

The decade of the 1970s was the golden era of old-growth research: careful descriptive work in the Andrews Forest quietly documented the size, species, and ages of trees, the diversity of plants and animals, and the ecological processes characteristic of old forests. Forest ecologists created "stem maps," inventories of all the trees living and dead, standing and lying on the ground in plots one hundred meters on a side, and then tracked growth, death, and toppling of trees over the decades within those plots, located in a wide range of temperature and moisture environments. Analysis of plots placed in forest stands ranging from very young to more than five hundred years old provided insights into changes in species composition and stand structure as the forest ages over successional trajectories spanning centuries. Using rope-assisted, rock-climbing techniques, scientists ascended into the forest canopy to sample tree architecture and investigate canopy-dwelling invertebrates, small mammals, and epiphytes such as mosses and lichens. Other scientists dug into the belowground ecosystem, with its tangle of roots, webs of fungal filaments, and rich microbial and invertebrate communities of cryptic creatures.

This research disclosed that the forest ecosystem is far more marvelously intricate than anyone imagined: the forest was revealed to be much more than its trees. The diversity of ecosystem components is mind-boggling—more than four thousand species of invertebrates and hundreds of species of lichens inhabit the forest. Species of microbes number in the many thousands. But surprisingly, the number of dominant tree species is just a handful: Douglas-fir, western hemlock, western redcedar, Pacific yew, and several maple and true fir species. Some previously unremarked species drew special attention: the lettucelike canopy lichen *Lobaria oregana*, for example, "fixes" atmospheric nitrogen, making this vital nutrient available to support the growth of the massive, ancient trees and to feed the ecosystem as a whole.

Ecological sciences have provided new conceptual frameworks for making sense of the complexity inherent in natural systems such as the old-growth forest under close scrutiny at the Andrews. These frameworks have included insights into how living spaces within the forest—niches—are partitioned by the availability of light, nutrients, water, and even sound. While the food webs that link all species are intricately tangled, the question of who eats whom can be usefully simplified into an understanding of trophic structure within ecosystems. Biogeochemical cycles depict the stocks and flows of nutrient elements and carbon throughout living and dead parts of an ecosystem. The concept of disturbance regimes—the frequency, severity, and spatial pattern of disturbances across a landscape and over time—provides multiscale views of forest dynamics. Understanding the roles of disturbance and death complements the centuries-scale change of living parts of the ecosystem viewed through the lens of forest succession. Life and death, order and chaos, continuity and change, all intertwine in the ongoing procession of an old-growth forest. Despite decades of inquiry and countless pioneering discoveries, the study of the forest constantly generates questions faster than we can find answers.

The trajectory of old-growth forests as a subject of science investigation and as a nexus of societal engagement with forests is displayed in the juxtaposition of two publications with roots in the Andrews. The research conducted in the 1970s culminated in a synthesis paper (Franklin et al. 1981) describing what was known of the ecology of old growth at that time. This rather simple characterization of the complexity of old forests, illustrated with stem maps, forest structure diagrams, and photographs, set the stage for the battle over the future of the forest that raged in the late 1980s and early 1990s. After nearly three decades of intense, lingering dispute, a new synthesis (Spies and Duncan 2009) portrayed attitudes and reflections of veterans of the "Forest Wars," including environmentalists, timber industry folks, ecologists, social scientists, and a philosopher. The societal context seems ever more fraught, even as the forests grow into their sixth century.

Each Step an Entry

LINDA HOGAN

Beneath us this place is seething with growth, the microfilaments of life passing around and across one another, an entire force that is feeding the future. It is the elder forest, the broken, the old, the almost past, where even what is dead continues to feed others. The feeling of life here is palpable, and the next generations are growing. The forest is moving with spiders and insects, although they are not as visible as they were on a day when I took the wrong road, 1501, where the surface was so burned I could see the different species trying to weave the top world together, seaming it together, in the same way that life below is doing even now. Here, there is great energy, and it is only a body sensation transmitted from the earth body to mine in our own symbiosis. It is in the soil. It is in the mosses.

There must be millions of life-forms in the soil working in darkness to create. More than millions, in truth. Everything is being taken up by everything else and excreted. A human can stand here and feel the great world at work beneath. It is nitrogen being given back, a place of offerings to the young.

I went to the water; I know it has great life, but it rushes so quickly to where it is going, and I don't like the fastness of it any more than I like fast people. I see the work it has done, smoothing stones, creating a valley. I do love the water when I discover a spring coming from deep in the earth, gently rising, and washing downhill. Or coming out of the stone, as it does at home. Running down the face of rock, shining as it moves to another world from inside and beneath.

The birds are present, and animals move in dry leaves, and I recall once hearing an owl in daytime. Each step taken into the forest is an entry into human silence. Each part of the old-growth forest is as amazing and resonant as any holy site.

Cosymbionts

VICKI GRAHAM

From the larger poem *Debris*

Geologist. Hydrologist. Botanist.
Saprophyte. Raptor. Arthropod.
A watershed broken into pieces
by specialists.

Like science, poetry is an art
of dissection—it is the tiniest part
the poet wants—fern spore, leaf pore,
bud scar, the veins of an insect wing
catching the sun, the barbs and rachis
of a swallow's feather in flight.

Like science, like poetry, love
for a place is an art of dissection.
The fingertip strokes the smooth pockets
of *Lobaria oregana*. The eye
sinks into the calypso orchid's silks.
And at dusk, the last note of a thrush
trembles in the ear.

Forest debris, from the French *debriser*:
to break into pieces—
a creek stone   a curl of moss
a thread of lichen   a spotted owl
a coralroot   log jam   earth flow

a cedar splitting in half
as the crack under its roots widens—
analysis of a circle
can begin anywhere:

with the fir fallen after a storm
and the slow decay of bark and heart—
habitat for fungi, nurse log for new saplings,
shelter for voles;

with the red-backed vole
gathering fir seeds, digging truffles,
leaving droppings rich in fungal spores;

with the fir seed sprouting, root hairs reaching
through rotting wood, meeting mycelium,
tapping the mycorrhizal mat;

with the fungal hyphae wrapping the fir's roots,
exchanging water and nutrients
for carbohydrates.

Fallen log, red-backed vole,
truffles, mycorrhizae—
how else, except by breaking down,
can a researcher understand a biosystem?
How else, except by keeping whole
through breaking down, can a forest grow?

The Art of Science

VICKI GRAHAM

From the larger poem *Debris*

Listen to the sounds of the forest,
the sounds of the land, the sounds
of a place loved and touched
by human hands. A boulder rolling down
a hillside, stone nudging stone in creek water,
the shudder of an earthquake.
Trill, whistle, hum, grunt,
growl, chatter, coo. Pods snapping
in summer heat, beetles
chewing rotting bark. Bear bells,
laughter, the grinding of gears
as a log truck winds down the mountain,
a whispered promise.
Geophony. Biophony. Anthrophony.
A forest symphony.

Let the words of science be brittle
as yew bark, sticky as monkey flower
three-pronged as a fir cone—
let the words of science be honed by use,
shaped by lips and tongue,
rounded, smoothed, cut, faceted
as stones tumbled in creek water
in winter floods—feathery
as moss silked with rain,
papery as the lobes of lichen

fallen to the forest floor,
spongy as an old-growth log
slowly leaching nitrogen into the soil—
let the words of science be things
and the imagination play with etymologies:
mycorrhiza: fungus root
cotyledon: kotyle: cup
plicata: pleated
let the tongue linger over *involucra*,
fascicle, bract, the teeth crunch on
clastic, schistocity, calyx.

Begin with love, a composite
of science, art, imagination,
and the pure world of the senses—
with the things the hand can touch:
the peeling papery bark of a yew,
the curve of a snail shell,
the beard lichen's wiry hair—
with color and taste and smell
and the call of a thrush at dusk,
the velvet glide of a spotted owl.

Begin with love and the questions
the heart asks: why stones
arrange themselves in lapped patterns
on a gravel bar. How old-growth trees
resist the quickly mutating pathogens
that attack them. Where the orchid
lacking chlorophyll gets its sugars.
Let the body then the heart learn
the forest and remember:
Data collection, computer analysis,
digitalized imaging begin
with hand and eye, tongue, nose, and ear.

And while the pencil hovers
over the page or the hand grips
a water gauge, the scientist
has time to stroke the willow leaf's silk,
breathe in the lemon scent
of chanterelles, follow the arc
of a swallow's flight.
The artist, too, has time to taste and touch,
and then to study the moss spore's journey
from protonema to gametophyte, time
to trace root and hypha to fir
and fungus, count
the fir cone's three-pronged bracts.

From
Drainage Basin, Lookout Creek

VICKI GRAHAM

From the larger poem *Debris*

In an old-growth forest,
not even the earth is still:
stakes set four square
trace trapezoids twenty years later.

A science of debris:
a bit of bark, a scrap of lichen,
a dropped needle or cone,
a stone shifting on the gravel bar,
a shed snake skin, a wasp
depositing eggs near the wood borer's larva—
pages torn from a book—read by a research.

Each stake stumbled upon in the middle of the forest,
each aluminum tag and magenta flag,
each rope reaching up into the canopy
or pipe reaching down into the creek
tells the walker: Someone has measured
this flow, tested this soil, weighed this log,
tells the walker: Someone loves this place.

Log Decomposition

JOAN MALOOF

The dead in a real forest belong,
they are beautiful there.
They die in each other's arms,
or their bones shatter,
as they hit the ground.

Or their lumps have become
so rancid and strange
that you are not sorry
to see them go.

There are no two deaths alike
when they come at their own time.
When a life is over, then,
all that's left is light.

In a real forest trees do not have wounds
straight-lined like surgery.
The dead here have been murdered
and lie like corpses in a mass grave.

The clues are plastics and metals
in shades that don't belong.
The victims, cut at the ankles
and laid at the feet of the living ones.

Those left standing cannot run or turn away.
Mosses cover the bodies with a blanket
of green, out of respect, but the trees
can only drop needles and seeds.

The clothed apes have visited the bones
year after year, discussing,
their elegant experiment.
But the study will never be over, not even then.

Decomposition and Memory

AARON M. ELLISON

First impressions—the head is enthralled, the heart is repulsed.

There is something interesting and exciting going on here. Ecologists—members of my tribe—were here! I revel in the familiar: tall and wide plastic collars are set to measure log respiration—the loss of CO_2 from these decaying logs—and plastic funnels move water running off the moss-covered log into plastic carboys that collect the throughfall for future analysis.

How do decaying logs contribute to the carbon balance of the forest?

What are these tubular excrescences? Are they part of the log? Melting into the log? No, just silicon skirts connecting them to the log. Why is one pipe erect, and the other recumbent? Logs decay after centuries, a plastic cup takes at least 250 years to decay; how long will the pipe tube cylinders last? Across the path, a plastic funnel and a plastic bucket (no, not a bucket, a *carboy*) broken; shards on the forest floor becoming the nanoparticles of the future soil. Will they end up in some earthworm, some mushroom, some newt? This installation violates the integrity of the forest.

There is wrongness here.

Second thoughts—what has been learned?

I know this experiment. It is a canonical example of a truly long-term study. Established in 1985, intended to run for two hundred years, it will reveal elements of the process of decomposition heretofore unappreciated by ecologists and foresters. The experiment

was designed thoughtfully, with careful attention to replication, sampling interval, and analysis. I explore the archive of the H. J. Andrews Long-Term Ecological Research site; there are scant publications on this experiment, and then only from the first few years after it was established. Data are posted online, in some cases through 2001, in others only through 1988. Available data on log respiration measurements run through 1995 and have never been published. Are data still being collected from the respiration tubes; looked at and analyzed; prepared for publication? How fast are these logs decaying?

These logs are impermanent. The forest is impermanent. This world is impermanent. Are any parts of the log permanent—the molecules, the atoms? The carbon in these logs was in the atmosphere, was fixed by the trees when they were still alive and used to build cell walls, trunks, branches, leaves. This carbon is now being returned to the soil, to the water, and back to the air. It is respired by the log, it is ingested by the mushroom and the flies, and it is reinspired by the trees around us.

It is enough to know this; I do not need to know how much or how fast.

The next day—what we remember, what we forget.

What is remembered in a day, much less after two hundred years? The log decays, and as figure 1.06—a modification of Michener et al. (1997)—shows, we envision a simultaneous decay of the integrity of the experiment.

The log decomposition experiment is not just the data. Recalling specific details, juggling idiosyncrasies of methods (in fact, the methods for measuring respiration changed in 1992), and ensuring validity of future measurements all are contingent on maintaining the experimental apparatus. The collars are only weakly attached. Silicon plugs (for temperature probes?) are exposed, extruded, or fallen. Is this experiment maintained or unmaintained? Does it matter, when the sampling interval is eight years, sixteen years,

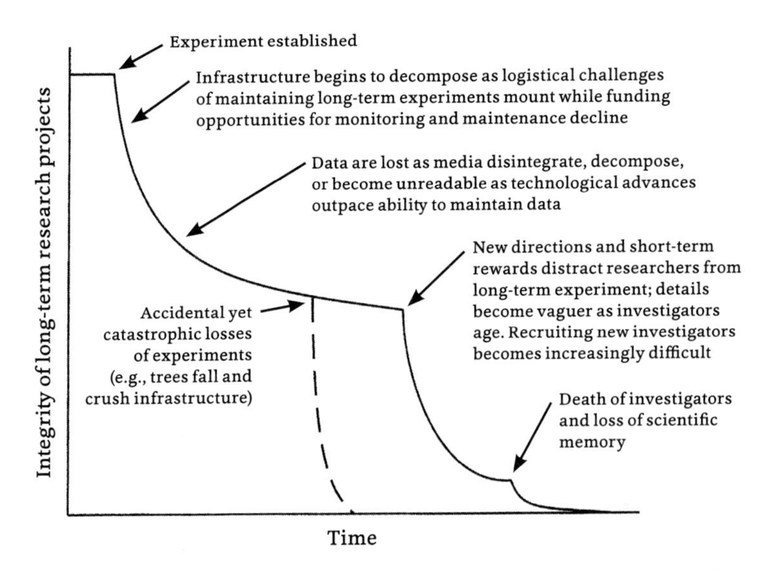

Figure 1.06. From William K. Michener, James A. Brunt, John J. Helly, Thomas B. Kirchner, and Susan G. Stafford, "Nongeospatial Metadata for the Ecological Sciences," *Ecological Applications* 7, no. 1 (1997): 330–42, http://lits.bio.ic.ac.uk:8080/litsproject/Michaeneretal1997.pdf.

two thousand years? Maintenance becomes more critical as sampling interval lengthens, but it also becomes easier to put off maintenance for another season.

It is similarly difficult to maintain the integrity of the data themselves—numbers on paper, magnetic bits, optical codes, droplets in the cloud of the World Wide Web. The half-life of storage media is falling as quickly as new media are invented.

And the interpretation and publication of the data are themselves data. If the data are not published and interpreted, will future generations of scientists know they exist? Will twenty-second-century graduate students be inspired to continue maintaining this experiment, measuring tissue-mass loss, log respiration, throughfall chemistry? The broader issue: How do we maintain continuity of research and researchers? How do we maintain enthusiasm for ecology, for this science? Do ideas and intellectual fashions decompose, too, only to be reborn in a new experiment, a new subdiscipline, a new journal?

* * *

Science is narrative, and these decomposing logs tell a story. But even though only twenty-five years have passed, the story is already fragmenting. Like the monks poring over the relic of Saint Leibowitz, what do we now think of these logs' original state—8.9% outer bark, 4.0% inner bark, 28.5% sapwood, and 58.6% heartwood—and what will we think two hundred years from now? I can see no outer bark, and I don't know if it even mattered. I don't even know how to read the story of the bark. The logs remind me of the stone walls of my New England home. The forests were cleared; the walls were built, and they bounded and defined the lives of the early European colonists. The colonists moved on; their stories are in fragments; and like these fallen logs, the stone walls meander through the now-regrown forest.

Boulders erode, logs decay. We know the "big picture"—cellar holes, pasture walls, boundary walls—but the details are lost: whose house, what was he thinking the morning he laid the first stone or placed the last?

More fragments.

The cost—what is this knowledge worth?

The scientist in me learns the details: 530 logs, 5.5 meters per log, each 0.5 meters in diameter, removed from intact stands and clear-cuts along the 1506–630, 1506–320, 1506–350, and 1506–354 roads. I calculate: 182.1875 cubic meters of wood, or about 50 cords. Enough to heat the average-size, modestly insulated New England home for ten winters. Access roads into old-growth forests were built simply to haul in and emplace the logs (damage was minimized); after the logs were entombed, the roads were closed, pyramids resealed. Logs were measured, experiments were established, data were collected.

So many trees, so much wood. Clear-cuts rending the forest in the service of science.

The passive voice astonishes. Who was responsible? Who built the roads? Who gave priority to saving old-growth trees over saving small trees? How did he feel when the big trees came down, were bucked into pieces, were yarded into place?

All to learn that logs decay.

I drive the 1506 road through the old growth, past the regrowing clear-cuts. Without a map, without a guide, it would be impossible to know where the logs came from, where their final resting place is.

The logs decay, the trees inspire, the forest returns.

GROUND WORK

Decomposition

As a topic for scientific research, decomposition is not very sexy. The ecosystem science world is dominated by "life scientists"; death and moldering are harder to get funded. Yet decomposition processes are an essential step in the cycling of carbon and nutrients. Mark Harmon has spent a career on the topic. As the "head rotter" among Andrews Forest scientists, Harmon employs a vast array of experimental, observational, and modeling techniques, generating data that feed sophisticated models used to compute carbon dynamics under different forest growth, management, and climate change scenarios.

Recognizing the importance of dead wood and decay in ecosystem and forest management, Harmon coined the term *morticulture* to identify management practices relevant to culturing the presence and functions of dead wood in ecosystems. Morticulture complements silviculture, the study and management of the living-tree component of forests. Given the many vital ecological functions of dead wood—habitat for many animal species, substrate for plant growth, source of soil organic matter, component of forest hydrology—Harmon contends that we should be as thoughtful in the management of the dead as we are of the living trees.

Harmon established his log decomposition experiment in 1985, and it has become a monument to the spirit and consequences of long-term ecological research. Previous to Harmon's work, scientists conceived of log decomposition as a slow, rather simple process proceeding over centuries; decades of detailed research reveal the logs to be complex ecosystems brimming with life and chemical processes. This landmark experiment set in a quiet patch of old-growth forest involves hundreds of logs of four species of conifers representing varying degrees of decomposability. Sections of large-diameter PVC pipe glued to the logs' flanks serve as respiration chambers to sample CO_2 exhalations from the decomposing logs. Subsets of the logs are sampled on a schedule extending over two hundred years, the expected lifetime of the logs. The sampling involves cutting cross-sections, or "cookies," from the logs and then analyzing chunks of the cookies for nitrogen and carbon content, invertebrates, zones of rot, and much more. Other decomposition experiments at the Andrews Forest examine decay of limbs and needles on land, of logs, of leaves in streams, and of roots in the soil, which are all part of deciphering the larger story of carbon dynamics. Understanding this system, involving all the storage sites and flows of carbon into, within, and out of the forest, is a foundation for sustaining forest productivity, achieving desired levels of carbon sequestration, and meeting other objectives that society may impose on the forest.

These studies confirm old stories and offer fresh insights. Depending on wood chemistry, common species of trees have a wide range of decay rates: cedar is very resistant to decay, but you can stick your finger into a Pacific silver fir log after only two decades of decay. Harmon was surprised to learn that nitrogen fixation occurs in decomposing wood at a low but significant rate; that a decaying log contains more living cells per unit volume than the heartwood of a live tree; and that fungi pull nutrients from the logs, incorporate them in mushrooms on the log surface, which then tumble to the forest floor. A big-picture story is that dead wood constitutes a major component of the carbon sequestered in native forests, an important consideration in morticulture practices intended to retain carbon on the landscape.

In the Experimental Forest

ROBERT MICHAEL PYLE

And here is what the scientists see
but cannot say:

How the dogwood blossoms glow
against the black wet trunks of Douglas-fir;
how the skin of yew runs red in rain, the bark
of young vine maple green as skin of anole
in a hot southern wood.

The way evergreen violets erupt hot yellow
from the green magma of moss, and trilliums pink
out, paste their petals to the waxy leather of salal.

The manner in which Douglas squirrels inscribe
the snow, and where they leave their middens.

Cascara's small tongues lapping the drip
as chorus frogs and winter wrens sound
the walls and depths of Lookout Creek. Pipsissewa
and bunchberry catching all the windthrow
winter can bring. All these things

may have adaptive value, for all we know.
Could generate data, yield understanding,
render the answers that poets may dream
but cannot write.

As last year's bracken rots beneath the new sword ferns
and varied thrushes whistle through spit,

I have faith
that somebody, somewhere,
surely knows
what to make of all this.

Notes for a Prose Poem

Scientific Questions One Could Ask

ROBERT MICHAEL PYLE

1. Sunbeams slanting through the forest strike evergreen violets and tight buds of cherries: what is the ignition point of each?

2. What is the relative albedo of snowmelt trilliums, rain-wet Oregon grape glister, and the pale underside of *Lobaria* lichen against cedar frond?

3. What is the precise incidence of sunshine that makes one centimeter of one web shine emerald, a nearby centimeter of another strand sapphire? Or waterdrops on Douglas-fir needle tips on sunny mornings after rain or snowmelt: one ruby, one tourmaline?

4. Is it innate or learned behavior that causes some birds to respond responsibly to pishing, outing themselves for a look, while others obstinately hang back, seeping and teasing invisibly from the brush, for as long as the investigator keeps pishing?

5. What is the capacity of the winter wren's heart for ebullience? How much blood must it pump for one endless obligato? How many times must a winter wren sing, to establish his territory for good? To get a mate? To achieve transcendence in the human heart?

Among the Douglas-Firs

JOSEPH BRUCHAC

Why is it
that one tree dances
while another,
mere feet away,
stands still?

Is it something more
than the vagaries of wind,
the differing shapes
of their branches?

Is it the way
their spray of needles,
like outstretched palms
cup the breeze?

Or is it just
that just like us
some have
at one time
or another
more of
an inclination
to move
to a rhythm
all their own?

From *Where the Forests Breathe*

BRIAN TURNER

Nobody knows how little we know
  about this forest. And nobody
knows how much time we have
  to piece it all together either,

nor how many mistakes we can make
  and survive. So best believe
the ineffable gives life to what we
  can love and revere, as when

we revel in the vine maple's red riot
  in new-growth forest, and marvel
at the gleaming porcelain shine of
  mushrooms piquant on mossy trails.

And here, then there, along
  Lookout Creek, golden maple leaves
parachute down, their descent
  a rhythmic, slow-motion dance.

From
Varieties of Attentiveness

FREEMAN HOUSE

After a good deal of climbing and crawling, I find what I'm looking for, a spot in the sun with a view of the creek, all bright white water here curiously dimmed by the banks of snow at its edge. Every fiber of me wants to sink mindlessly into the respiration of the life around me, and I do so until my breath slows to some approximation of the breath surrounding. A variation in the sound of the water snaps me alert. I need to pay attention or I'll have nothing to write about. Bummer. Shall I count trees? Not a chance. Shall I name species? Too long since my tree identification class at what was then Oregon State College. I decide instead to speculate what time it is in the scheme of natural succession at this place. The largest trees are Douglas-fir, I'll guess three hundred or more years old. I look for Doug-fir seedlings and see very few, but instead see a multitude of hemlock seedlings sprouting straight out of rotten wood. I lie on my back and look up, an action that threatens to put me back into my preferred trance. Focus: The highest and most abundant crowns are Doug-fir, maybe 250 feet up. But wherever there is some opening in the canopy, I can see western redcedars and western hemlocks struggling upward, occasionally to within a few score feet of the highest canopy. They are tall but not vigorous looking. They are waiting their chance, once the giants beside them fall and let in the light, for both themselves and their sprouting offspring below. Another five hundred years, they may come to be the boss trees here. My rational mind tires quickly these days and I retire back into mindlessness. Spring manifests in an occasional cloud of no-see-ums. There is a breeze in the canopy that doesn't reach the ground but knocks down a fine orange mist of what I take to be pollen. After a time without time, I rise and look around for a way back. I find that by climbing straight up the slope I will reach the trail that after all is only a hundred yards away.

Attention is not the same as attentiveness in the sense that I would like to use the word. Attention is narrowly focused and intense; it can be maintained only for relatively short periods of time, although it can be made cumulative through records stored in databases. My small efforts evaporate as soon as I stand. They won't stay with me unless I return again and again. To be attentive in the sense I mean is to work toward becoming a functional part of a place. To begin to be attentive to this little patch of space, I would have needed to return here again and again over the last twenty or so years. I would have needed to bring my children with me and had conversations with them about the place. What does it mean to live here? What are our responsibilities? What can we take from here to nourish ourselves without harming the place?

Despite a culture that encourages nonattachment to a particular community, I would speculate that attentiveness can lead to reconnection, to communion with places. Attentiveness is a personal and necessarily long-term practice; it can quickly become a community practice. In my home place, the practice has taken the form of the contemporary community going out to engage the landscape we inhabit and attempt to repair some of the wounds that we and our ancestors may have inflicted on the place. After twenty-five years, that communal practice has become one of the signifiers of our local community. Science has always played a role: the new science of old-growth Douglas-fir ecology that has been generated at Andrews Forest has played an important role in helping us defend our own ancient Douglas-fir forests. But it's hard for me to see how scientific method by itself can get us to that desired state of belonging, a state of being that includes an intuitive sense of how to act so as to do no harm. You can't immerse yourself in a landscape by translating the landscape and yourself into numbers and formulae. On the other hand, I am convinced that my community's style of attentiveness and that of Andrew Experimental Forest have more in common than they have differences.

A world without H. J. Andrews Forest and the Long-Term Ecological Research project would be a poorer place. It would be enough to see the aerial photo of the Lookout Creek watershed surrounded by ubiquitous Forest Service clear-cuts to convince me of this. Knowing something of the work that goes on here elevates the place, in my mind, to something that approaches the sacred. In the generation during which Andrews has been a site for study, it has developed resemblances to deep ecologist Paul Shepard's description of belonging. Some of the scientists with whom I've had the good fortune to spend time have without a doubt become a part of the place. They are attentive. Their enthusiasm when describing the minutiae of their work knows no bounds, including the one that limits my capacity for absorbing so much new information in a short time. They have a poetic quality; they love this place and their work very much. Their work has changed their behavior. But when they retire, what will they leave behind but a cold but thorough database on which to build?

When I arrived here, I spent the day guided from site to site by the same man whose enthusiasm knows no bounds, and so forth, and at the end of the day I blurted out the question "When will all these data be interpreted? I want a picture of a watershed as a living being!" He didn't miss a beat but, with a twinkle in his eye, answered, "Don't know; we're all here as dispassionate observers." I've been puzzled by that reply all the time I've been editing these notes. Even remembering the spark in my friend's eye, it has taken me this long to realize that his very being—his enthusiasm and sense of vital participation—is a living form of the integration I seek.

Aquatic ecologist Jim Sedell writes in *In the Blast Zone* of what he calls the tribalism of scientific endeavor, a quality most likely to be generated, he believes, when organized around a single place. Unsurprisingly, Sedell first discovered said tribalism as a postdoc at Andrews Forest. "Communities of scientists sometimes coalesce into these intense group endeavors. I think of them as science 'tribes,' gatherings of motivated individuals sharing an excitement of discovery and personal dedication to solid research heightened by a sense of shared purpose, rich camaraderie, and, I can only call it, *joie de vivre*." In other words, they become people whose sense of self is reconfigured in the context of a particular place, people whose behavior is transformed.

It is the institutional memory of these transformative experiences that will make cold databases come alive. Such a conclusion is, of course, contrary to the theories and practices of Science. But in two hundred years, the forested watershed of Lookout Creek is not the only thing that will change. Could it be that out of the turbulence of the times that coincide with the Long-Term Ecological Research, all those atomized data will appear suddenly emergent as a living whole? We can only wait and witness.

Poetry-Science Gratitude Duet

ALISON HAWTHORNE DEMING
FREDERICK J. SWANSON

PATIENCE

Swanson: I thank you, as a poet, for the patience that you display in your engagement with the forest. I've been in the forest with poets, and when I've said, "Let's go," they have said, "I'd like to stay here three or four hours more—maybe a poem will happen." I love that patience to stay, watch, listen, be open to new possibilities. I'm concerned that scientists have become impatient—earnestly collecting data to test hypotheses but failing to let the ecosystem speak to them beyond the bounds of their immediate task. Growing commitment to electronic sensor systems and massive data streams further removes us from opportunities to be open to the natural world in our work.

Deming: Well, thank you for showing me the patience of the scientist. I feel gratitude to witness how a scientist will doggedly chase a study and phenomenon—even over decades. A landscape can look static to an untrained eye. But slowing down to see patterns of disturbance, recovery, earthflow, to understand that it takes two hundred years to know how a dead tree might influence its environs—these speak to me of a dedication to seeing the unseen; and that is a very poetic way of being. Thank you for showing me that while I flit through the landscape like a hummingbird gathering nectar, if I took direction from the methods of science I might ask a question that would merit a long, long view.

WORDCRAFT, VOCABULARY, STORY

Swanson: I so much admire your skills with language—your vocabulary and ability to put words together, sometimes in uncommon combinations with interesting new effect. You can say what we scientists may feel but cannot articulate, or that we do not even realize that we feel. What skills we once had for expression of emotions and out-of-the-box description have long been professionally scrubbed from our minds. Most scientists must make a conscious effort to break the mind and tongue loose. For science to benefit society, the findings must be delivered in good stories; thank you for sharing your gift for storytelling.

Deming: Your precision with scientific language moves me, the way it constantly cooks up new words filled with music and metaphor. I am grateful for "morticulture," "cutthroat fry," "disturbance regime," "old growth," and yes, even those names given that seem inadequate to spur the imagination because they invite me to go for new metaphors. I tell my students to be specific not generic in their naming. Science does just that in its imperative for accuracy. But science does not always see the power of its own metaphors. I long to free scientific discourse from its utilitarian yoke and see what can sing in its vocabulary.

WONDER, ASTONISHMENT

Deming: As writer Kathleen Dean Moore, our friend and collaborator, says, "You feed our astonishment," and I thank you for that. It's weird to me that empirical findings can spur that transcendent sense we call wonder. But certain data can verify for me the sense that natural systems in their beautiful complexity have a great story to tell, perhaps the greatest story. How did we come to be here? Where are we destined as a species to go? I felt this so keenly when I wrote about monarch butterflies, who would have remained a sentimental trope to me had science not taught me how brilliant these organisms are in their capacity for migration and metamorphosis. Nothing excites me like the story of how life has articulated itself on this ball of star stuff that is our planetary home. The more one learns about this, the more astonishing it all becomes. And the stronger becomes my sense of the imperative to use language to see and honor it.

Swanson: Thank you for your expressions of wonder and astonishment. We scientists have these possibilities, too, but seem to get a bit calloused by familiarity with our subject matter. The example of your work—and your spirit—reawakens the sense of astonishment in us. I am confident that scientists have a sense of wonder that is heightened as we plunge deeper into the phenomena we study so intently; but for many of us, it is difficult to convey that sense. The leap to a poetic view of the world can help us break loose and tell our stories with more passion.

FAITH

Deming: I want to say something about faith, not as a religious principle, but as an informed relationship to natural process. Thank you for showing me that science has faith in the material laws of process and becoming that define not only the outer world that so fascinates us but also the inner one that drives our curiosity and sense of belonging to something much larger than our individual selves.

Swanson: The faith of writers in their work so impresses me. Scientists can have a sense of being grounded in data collection, in the use of instruments, in analysis, and in tightly referenced publication in journals. Writers rely on their own wits and a faith that their stories will find and influence readers whom the writer will never know. The compulsion of writers, fueled by that faith, shapes how we all view the world.

EMPATHY

Swanson: Poets display such empathy for other living creatures, both animal and plant. The depth of scientists' relationships with the natural world certainly reaches that point, but we seldom speak about it. The blend of empathy and mystery draws us all to the forest. In her book *Braiding Sweetgrass*, our friend and writer-scientist Robin Kimmerer refers to this as "seeking kinship." It's the writer in her that brings the feeling forward.

STILLNESS

Deming: Thank you for being calm and still in the face of vexing challenges, for being like a mountain. Sometimes you seem as quiet as stone, and yet I know you are present to a sense of time that eludes the artists in their haste to make something new. Maybe we need to make something old and very still in the face of the velocity our kind has created.

CULTURE

Swanson: Thank you for helping me see the cultural dimensions of our work—the work of scientists working alone, and our work as collaborating teams of scientists, artists, and humanists. I realize our work at the Andrews Forest has helped the public with ways of viewing iconic features of our Pacific Northwest landscapes: ancient forests; cold, fast rivers; salmon; and volcanoes. Our work together greatly amplifies this form of citizenship, helping fellow citizens of the region understand these landscapes, what they do for us, what we must do for them, all in a spirit of respect and reciprocity.

SKEPTICISM

Deming: I tell my creative writing students to anticipate the skeptical reader. We are all skeptics and should be. Science has taught us this, and I am grateful to science for a method that finds in each answer a new set of questions. This curiosity and eagerness to learn from what we learn seems acutely necessary to our times, when so many human choices have such powerful planetary consequences.

HOPE

Deming: I value the curiosity and eagerness in our partnership and thank you for embracing the simple question we seem to be asking: what would happen if we brought our often polarized ways of knowing together into one big commitment to the planet that has spawned us? "What do you know?" I want to ask my scientist partners. "How do you ask your questions? What do you do with the answers you get? Why does this work matter? How can we pass on this vessel of embers to those who come after?"

Swanson: And I want to ask these questions of you poets. We both respect and wonder at one another's way of engaging with the world, and we share a love of the natural world and a story well told. Our work together expresses our hope for the future.

– Part Two –

CHANGE AND CONTINUITY

Genesis

Primeval Rivers and Forests

PATTIANN ROGERS

If these weren't so very ancient,
they might easily be found. But they are
deeper than subterranean Siberia,
of a longer past than the oldest lichen
fossil discovered in Rhynie soil, from
farther away than found meteorite
remnants of three billion years.

These primeval forests and rivers
were the first to believe in trees dead
but standing. They were the first
to envision the living in the decay
of the down-dead, the first to conceive
possible orange rills of fungi, fluted
white helvella, beetles, spider mites
and spotted newts, a warty jumping slug
hidden beneath fallen needles and duff.

Birds were among them then before
there were birds, being mere wings of sun
off the rivers before there were rivers,
being mere flitting shadows in the upper
canopy before there were shadows before
there were canopies of flitting leaves.

And although these ancient waters
flowing through storied rain forests
have never been told, I imagine how
they imagined before they conceived
fish as smooth as silver glass, fat
and buoyant on river bottoms, how
they dreamed those fish swirling
in schools of crystal to the surface
without yet having bones, with no
eyes of gold or scarlet gills, before
flood or drought, current or cutbank.

Today the hiss of a single stem
of seeded grass alone in a slender
wind recalls the silence in far rivers
and forests preparing for themselves,
a silence expectant of wind, expectant
of seed. A brief fragrance passing now
suggests their beginning from absence,
the fragrance of the origin of fragrance,
damp oakmoss, sun on decay, the scent
of nostalgia for a thing I imagined
I knew before I knew.

Forests and People

a meandering reflection on changing relationships between forests and human culture

BILL YAKE

The wet-climate forests of the H. J. Andrews Experimental Forest are much like those near my home in the lowlands between the Olympic and Cascade Mountains of Washington State: a mix of trees that colonizes grounds cleared by windthrow, fire, and humans—first with red alder, Douglas-fir, western redcedar, and bigleaf maple. Then, with accumulating time and shelter, the forest welcomes shade lovers: western hemlock, grand fir, and Pacific yew. Below these thrive varied understories of salal, hazelnut, Oregon grape, sword fern, rhododendron, and scores of other herbs and shrubs. Branch-hung with nitrogen-fixing lichens, root-threaded with fungal mycelia, and boll-chiseled by pileated woodpeckers, these forests turn within the slow cycles of nutrients, debris conversion, and slumping nurse logs, and the quicker cycles of day, night, and the wheeling seasons of sun, rain, fog, wet snow, mist-soaked moss, and cascading streams.

The forest grows on the past and into the future.

* * *

It is November—with attendant drizzle, squall, and rare sun breaks. Along the trails, fungi fruit everywhere. Short-stemmed russulas—I think of them as goblets sprouted for goblins. Chanterelles with golden caps ridged beneath like fingerprints, some big as my outstretched hand. Slippery jacks—the less couth cousins of the bolete (porcini) royalty. Oyster mushrooms, sulfur tufts, a hedgehog or two. And a horde of their kin—some bright and intricate as coral, some gleaming like icy-white gelatin. It's been a wet fall with no hard freezes: a perfect year for mushrooms. During my stay, while attending a gathering of mycologists and fungiphiles, I learn that only 5 percent of the world's fungi are "known," which is to say that only one species in twenty has been formally described and named by taxonomists.

Autumn rains have been sweeping in from the Pacific for a month now, and these forests seem to be mostly water. The voiced speech and music—the polyphony of the forest—is river-sound below and wind above. Rivulets and drops of condensate add complex rhythms—the slow chaos of xylophonic wood-notes. Now and then, there is the talk of a single chorus frog, the scold of an agitated chickaree, or the gossip of a raven. Deep in the forest the dim green light could be that at the sea bottom, the watery light down at the holdfasts of kelp forests. Lichens are the forests' algae.

* * *

I'm spending only a moment here—ten days cooking, writing, and sleeping in windowed rooms at the headquarters compound, which is set on the flats near Lookout Creek, about one mile above its confluence with the Blue River. The compound stands on ground laid down maybe forty thousand, maybe one hundred thousand, years ago, when glaciers out of the high Cascades blocked the Blue River. This glacial dam impounded a temporary lake that reached well up the valleys of the Blue River and Lookout Creek. The foundations of my temporary home are dug into the terraced sediments from that old lake. Beyond and above the headquarters compound, forested ridges rise and fade into the scudding mist and fog.

* * *

Ambiguity resides everywhere in language. Even in everyday words. We forget to define our terms and conversations go awry. So before reflecting on the changing relationships between forests and human culture I'd like to think a little about our words, our terms:

Forest (n): a dense growth of trees, plants and undergrowth covering a large area. A word of contested, possibly Latin, origin. *Forest* came to mean the hunting grounds of royalty—closed off to "commoners." The modern usage, especially in North America, has meandered a ways from this root. The alternative signifiers *wood* and *woodland* sprouted from the Old English *wudu* and *wuduland*.

Change (v): to alter, to make different (also, to give and take reciprocally; to exchange; as to change places with another). Originally from the Celts, it meant "to bend" or "crook," becoming—later—"to barter." To *make change*. A word threaded with commercial implications.

Words change, including *change*. Nowadays change might mean "slow evolution" or something far more cataclysmic. Forests change—their species evolving and coevolving, their composition shifting toward (or lurching away from) climax, their parts and entireties leaning, quivering, proceeding, shifting, and, in extremis, collapsing or disappearing. Forests stacked onto logging trucks can be seen disappearing into commerce.

People and cultures *change* as well. Values and behaviors morph; populations wax and wane; generations learn and forget; their tools and machines sharpen, stultify, evolve, or fall into disuse.

Words themselves evolve, their meanings shifting subtly or abruptly. What will *forest* mean in 2203? Or *change* itself? Flux runs in all directions and the Reflections program is only a decade into its two-century effort of tracking forests, perceptions, attitudes, science, definitions, and language itself.

Cultural ecology (n): the study of human interaction with ecosystems to determine how nature influences and is influenced by human social organization and culture. Where *culture* is the learned patterns of behavior and thought that help a group adapt to its surroundings. And if this culture batters its surroundings rather than adapting to them? What then?

* * *

Despite our best efforts, however, definitions cannot pin forests down; they are too multilayered, interwoven, and subtly animated for that. And there is also the shear inadequacy of language.

Consider the multidimensional gradient between forest and not-forest. The forest is complex, heathen, other, overwhelming, sobering; it is both immediate and aloof. The not-forest is another infinity of possibilities. At what point do heedless forces (winds, fires, humans, tsunamis, volcanic and climactic cataclysms), through their assaults, convert a forest into not-forest—into woodlot, parkland, blowdown, kindling, beaver pond, quarry, dump, berry patch, burn, stump farm, tree farm, clear-cut, "forest practice," reservoir, highway, landing strip, or transmission-line right-of-way?

And when do these, left alone, cross the reversion line back into forest?

Because forests predate us, they certainly can appear, disappear, and thrive without us. Perhaps that is what *wild* means: life thriving without human attention.

* * *

Forests swallow. Warm, wet forests have enormous appetites. The temperate forests of the Pacific Northwest swallow sunlight and summer heat, mists and drenching rains, traffic noise and wind, the breath of animals and machines. An occasional hunter, lost hiker, or stranded motorist. These disappear into forest, feed it, making more temporal forest-body and forest-verb. Forests have swallowed whole civilizations. Mayan temples and Haida villages. Perhaps only wildfires, tsunamis, and stars in nuclear fusion have greater appetites. And human cultures, for cultures swallow as well. Among their various snacks: forests; among forests' snacks: cultures, or at least their remnants.

Here, near the Blue River, human cultures have encountered Cascade forests for at least 13,000 years. The earliest documented human presence in Oregon—determined from coprolites found in the Paisley Caves east of the Cascades—dates from 14,500 years before the present. Clovis points have been recovered from a possible hunting camp on a bench above the confluence of the Blue and McKenzie Rivers. This era coincides roughly with the Magdalenian artwork of the Cro-Magnon people found in caves, including France's Lascaux and Spain's Altamira.

Since the glacial retreat roughly 15,000 years ago, the forests of Oregon's Cascades have shifted with climatic changes in temperature and rainfall. The postglacial spruce have largely disappeared. Tree lines have shifted, and alpine habitats have fallen and risen. The present forests of the Lookout Creek catchment probably established themselves some 4,000 years ago. Roughly ten old-growth generations.

Through most of the intervening ages, human presence must have barely bruised the forest. Some meadows and berry fields were maintained or enlarged by purposeful burning; withes and poles were gathered; and trails—those running up to the ridgelines and over cols to the obsidian quarries at the base of North Sister—were blazed and minimally cleared.

Along the margins of the Pacific Coast, wedges, mauls, and adzes were wielded to build the small collections of longhouses that faced the sea and its winter storms. Man was another clever creature in small herds. Humans nibbled at the forest.

Then Europeans arrived from lands where they had largely beaten back and tamed the forests. They arrived with different powers, ideas, and diseases. With new flora, much of which was invasive. With gunpowder; cattle and sheep; saws, axes, nails, shakes, froes, and sawmills.

The newly dominant culture featured photographers and springboards. Newspaper. A book from a desert god. But also Young

Science gathering its power. On one hand, this culture had a general disdain and fear of darkness, deep forests, wolves, and mushrooms. On the other hand, there were explorers and naturalists with an intense curiosity about this "new world." A love of cash and status and unfettered access to lands that, back in Europe, would have belonged to royal overlords. In time, forests fell, farms rose, mountaintops were scraped into Appalachian valleys. But also, national parks and forests were established and protected. A strange new concept—lands held aside from development for their viewscapes, geysers, indigenous wildlife, forests, meadows; held aside to provide unobtrusive access by regular citizens.

This is not to imply that all native people disdained wealth and class. The Haida, the Chinook, and other clans of rich warriors, sea hunters, and traders along the Pacific Northwest coast remind us of ourselves in that respect. But they were not interested in leveling forests. The idea would have seemed, I suspect, absurd—that sort of destructive power belonged to cataclysmic floods and tsunamis, to avalanches shaken loose by rain on snow, to landslides triggered by the shuddering earth, to temperamental mountain spirits.

Transplanted Europeans elbowed their way into the provinces of the calamitous spirits. They took the coastal forests down first, then those of the lowest valleys—with log booms, skid roads, ox teams, splash dams, and steam donkeys. Eventually they rolled inland and hauled the forests out on the trestles of short-track logging railroads.

* * *

As outliers, remote cul-de-sacs, the forests of Lookout Creek were spared. For the quirks of history, economy, and technology, we can, in this case, be grateful. Great trees—centuries old—still stand on the steep ridge-flanks beyond my windows, veiled then revealed by transiting squalls.

What were the driving wheels, the motivations, for bringing down the primeval forests? Jobs certainly, but also control. Dominance. Desire for sunlight and wealth. Status. The book of the desert god sanctioned "dominion . . . over all the earth." Boss and lumberjack alike pitched themselves against dark forests stocked with rough beasts—bears, catamounts, and owls with their haunted midnight calls. Many died—chokermen, fallers, and high-climbers—killed and maimed by widow-makers and snapped rigging. But the forests blocked access to the ports of manifest destiny. And felling them paid well. It must have seemed provident.

* * *

But it occurs to me that this is all too pat. I recall that my mother, born and raised in sunburned Fresno and the Texas plains, used her mother's inheritance to buy the pine-wooded lot beside the home where I was raised—the little house my father built in Spokane. She dug up, potted, and transplanted grand fir and redcedar from north Idaho. The green, the shade, and the buffer between us and our nearest neighbors seemed to calm her. "We should enjoy the smell of soil, its feel," she said once while planting crocuses; "it won't be there in heaven."

In that place, those acts, those words, I think, the seeds of my forest appreciation were planted.

No culture is monolithic. We easily overlook half-concealed sources of sustenance and greening. From the groundwater hidden in aquifers and hyporheic zones to long-lived seeds buried under Saint Helens's snow and ash; from the texts left for decades unread in the back rooms of libraries and bookstores to the secrets half-buried in the subconscious: resilience and inspiration reside. This protected pocket of old, functioning forest—my present refuge—is an apt example.

Even landscapes as blasted and bleak as the posteruption slopes of Mount St. Helens revegetate. John Muir and Teddy Roosevelt somehow emerged from the bleak and blasted cultural, social, placer-mined and logged landscapes of the Gilded Age. But the seeds of Muir and Roosevelt's inspired imaginations must have germinated in landscapes far from the blast zones of development. Muir, Roosevelt, and their conservationist brothers and their sisters devised the radical concepts of national parks and forests, a circle of cause-and-effect in which inspiration-and-result were pretty much one and the same. Wildlands inspired and impelled the creation of refuges *for* wild habitats and *from* devastating exploitation and urban anxiety of exploited lands.

As evolved creatures, we are partnered in a long, evolutionary dance with sustaining habitat. Even the way our forward-peering, depth-perceiving eyes and nimble thumb-opposed hands grasp a forest branch for a walking stick—for balance—is part of that dance. Over the past two hundred years, the scope of social responsibilities and conscience has slowly broadened. Positive tribal influences have rebounded a bit. The green movement has taken serious root in the values and politics of many. The works of naturalists, a full range of biologists, forest scientists, and geomorphologists—all the inquiries of honest science—continue to refine our insights and inform our efforts to preserve and recover forests. The idea that these will be enough to stay or reverse the whelming impacts of population growth, species loss, and climate disruption may now seem doubtful, even naive. Still they remain sources of hope and potential healing as we peer apprehensively into the squall, heat haze, and human misbehavior that veil the oncoming centuries.

From *Out of Time*

SCOTT SLOVIC

Today, after I've spent several days visiting the gravel bar next to Lookout Creek and looking at it from different angles, I notice that the solidity of the stones, mud, logs, and tangle of alders, firs, and hemlocks feels startlingly contradictory to the actual newness of this landscape in the context of geological time. This gravel bar is an infant, yet to me it feels permanent. At the same time, there is a vibrant sense of upheaval here, as if the entire sweep of water, hillsides, and trees is a sort of waterfall, frozen in form during the brief moment when I'm here to witness it. My friend John Felstiner is fond of describing waterfalls as "flux taking form"—he's referring to the "paradoxical dynamic" in Western nature poetry of the past two centuries, and in nature itself, by which "raw energy can show design." The gravel bar also strikes me as "flux taking form," as I can see the future of this place written in the not-flooding creek and in the many tilting trees, pausing now en route back to earth. Just because we don't see actual change, our eyes being as temporary as the rest of our being, doesn't mean change isn't happening right before our eyes. This gravel bar, it occurs to me, is a rare clearing in the forest, a viewpoint—strangely like a clear-cut, but without the debris of tree limbs, the steepness, the stumpage, and with many more rocks, mossy with new life. A kingfisher zips past as I sit on my wet boulder, following the flow of the creek. An airplane groans overhead, unseen above the clouds, its sound competing with the rush of water—and then there is only the water.

Ten-Foot Gnarly Stick

JAMES BERTOLINO

I found this great stick on one of my
brief hikes in Oregon's Andrews Forest.
Its length is slightly arced, and it has
rich character. I comprehend and treasure it
as the arc of my life. My seventy years begin
at the blunt end, and it runs pretty straight
for about three feet. Then things get very interesting:
there are sudden curves, and protruding knobs,
and it never marks a straight line for more than
a few inches at a time. There are gouges,
dark spots, and places where it threatens
to split. But there is a steady progression until
about nine inches from the end, where it
makes a sharp, left-hand turn. What, I wonder,
can that mean?

Pondering

JAMES BERTOLINO

Looking with pleasure up a mountain stream, then following the path of water down through channels in the rocky streambed, I notice the flat area of a rock slab has a bowl cut into it, where the water seems to slow and swirl, as though it is taking time to consider its own flow, where it has been, and what it remembers. When that eddy rejoins the rush, it seems brighter, with heightened flecks of white.

In the Palace of Rot

THOMAS LOWE FLEISCHNER

Learned today of something so cool, so subversive
and literally subterranean:
Root Rot Mortality Patches.
Major forest gaps in coniferous old growth
caused by large-scale (dozens of feet across) underground patches of
    fungus that selects a
      specific tree species—here, often Douglas-fir.
They rot out the roots of the Doug-firs within this circumference; the trees
    then fall
(in random directions—a key field mark—rather than aligned, as occurs
    from windthrow).
Also, one can often see sick trees (dead crowns, etc.) around the circle's
    perimeter.
    The gap releases other species—often western hemlock here, or
    redcedar—to go to town
      (i.e., to sky).
A major shaping force in forest communities,
yet invisible
and to most forest ecologists, even,
unknown.

GROUND WORK

Disturbance

From a disturbance ecologist's perspective the world is decades of boredom punctuated by moments of chaos. And the chaos—wildfires, landslides, floods—is the most interesting part. Take the February 1996 flood. A few feet of fresh snow blanketed the Cascades as weather satellites showed a series of warm, wet "pineapple express" storms approaching from Hawaii. A few, stalwart researchers rushed to the Andrews for a close look at a big rain-on-snow flood. They were not disappointed. Arriving on the evening of February 6, they found debris flows beginning to course down some of the small, steep channels of the Andrews Forest. Big wads of logs cruised down Lookout Creek. By midday of February 7, the creek's main stem reached a discharge nearly a thousand times its late-summer low-flow level. As the floodwater rose, it picked up more logs delivered by the debris flows or fallen into the channel since the previous big flood, and carried them downstream. As Lookout Creek cut new channels and filled in some of the older ones, a thorough rearrangement of the stream and riparian zone was under way.

The flood taught scientists some surprising lessons. One was the importance of being present during the disturbance event, since some of the evidence of important events was ephemeral. For example, in some cases snow plowed to the roadside created dikes that routed water down roads, causing damage that would have been nearly impossible to interpret once the snow melted. Researchers witnessed how the flood used floating logs and rafts of wood to rip up thirty-year-old alder forests growing on gravel bars along the main channel. Sampling of fish, salamanders, and other aquatic organisms before and after the flood revealed that some species (for example, cutthroat trout) were little affected, because they could find refuge in slack water close to the stream banks. Other species, such as sculpin and dace, small fish that live among the streambed boulders—which researchers could hear clunking along beneath the turbid water—suffered high mortality.

An essential feature of long-term, place-based ecological research is the opportunity to be prepared with predisturbance data to interpret a big event, to possibly observe the event itself, and then to track some of the consequences of disturbance through the course of ecological succession over the long haul. Andrews Forest scientists have capitalized on these approaches in studies of many disturbance processes common in the forest—natural processes such as floods, wildfire, insect damage, wind, canopy loading by wet snow, fast and slow landslides, and human-imposed processes such as logging and road construction, to name a few. Investigating this diversity of processes requires a variety of study methods, determined in part by the tools available and the extent of history that can be reached with those tools. The simplest case is to impose disturbances experimentally, such as clear-cutting or prescribed fire, and track ecological responses over decades. This has been done at Andrews Forest since the 1950s in the context of experimental watershed and silviculture studies. Although it may take decades to witness a disturbance, responses to them in the form of vegetation succession play out over centuries. Another approach is to retrospectively study disturbance history of the past several centuries and even millennia, capitalizing on the record-keeping capacities of tree rings, landforms, char on tree bark, pollen and charcoal layers in ponds, and many other subtle, historical features in the forest.

The broad findings from these disturbance ecology studies have reshaped both our views of forest and stream dynamics and how we might go about managing these systems. First, it is important to change our tendency to view disturbances in only a negative way. There are winners and losers in every disturbance event;

the death of established organisms creates new niches and living space for others, commonly of other species. Second, the types of disturbance processes in even a small area, such as Andrews Forest, are amazingly diverse, and it follows that their effects on ecosystems are similarly varied. But studies find that, despite all this disturbance, ecosystems are resilient—even after severe impacts such as large floods or clear-cutting followed by burning of the logging slash. Many "biological legacies" in the form of surviving organisms and propagules (such as seed banks and rootstocks in the soil) from the predisturbance ecosystem are carried through into the postdisturbance ecosystem, potentially influencing the overall biological response to disturbance.

We now recognize that repeated disturbances, such as periodic wildfire, are critical influences on ecosystem development, patterns of forest age-classes across the landscape, and species evolution. This dynamic landscape view is a major shift from earlier concepts of stable and climax systems and the "fully regulated" managed forest. A dynamic-landscape view prompted Andrews Forest scientists and land-manager colleagues to use their understanding of historic disturbance regimes to guide plans for the frequency, severity, and spatial arrangement of forest-cutting patterns on lands where logging is permitted.

New Channel

JEFF FEARNSIDE

It must have been sublime:
trees two hundred feet tall
tossed like twigs into the torrent,
broken in two, in three, swept
downstream with a flow of debris
several meters thick.

I walk along Lookout Creek,
banks scoured high and steep,
channel crosscut by
a logjam like so much kindling,
a latticework of old growth
growing no more.

How many fish died—
the torrent sculpin and longnose dace, their eggs—
water skippers,
litter spiders and their eggs,
caddis flies and their larvae,
choked, bludgeoned, buried?

And yet,

in the buildup of silt
clogging formerly clear waters,
a stand of willows waves
gently, as if offering tea,

hospitable even
in humble new beginnings. Moss gorges itself

green on the mighty carcasses
of downed firs in which
insects and arthropods thrive, while fish relax
in splash pools behind arboreal bars.
Salamanders spawn in backwaters
formed by the blockage.

Why do we focus always
on the destruction
and not the regeneration?
We reach for tales
assuring us of immortality,
yet we refuse to read

the life right in front of us.

Slough, Decay, and the Odor of Soil

BILL YAKE

Trunks, once poised and upright, collapse toward
a two-century graduation into beetle and vapor,
moss, conk, and seedbed—their boles intermittently
chiseled by woodpeckers uncoiling their barbed tongues
and probing the grub-etched galleries within. Hibernacula.
Loosened bark. Sap and heartwood riddled with crawlways
where ants stalk wood-mining fungi, where inexorable
ant-infesting mycelia reciprocate. The odor of must,
cedar disintegrating through pungency to pulp and soil.
The plush, ripe scent of continuous integration.
What seemed solid, stains and softens decade by decade,
to be torn apart by bears after ants: the flavor
on their tongues that of dull sparks. All is relentlessly
hollowed, grain by grain, cell by cell, into sponge and grub
dust, salamander refuge, slug haven, frog shelter, and moss
—all deepening to opulent, pre-ultimate, humus and duff.

From *The Mountain Lion*

TIM FOX

As I work my way upward through the forest, my mind wanders back to 1990, my second summer in Oregon, when, at age nineteen, I was one of hundreds of seasonal field-workers hired by the Forest Service to call for northern spotted owls (*Strix occidentalis caurina*) in California, Oregon, and Washington.

The divisive owl war that gripped the nation was at its peak, and one of the sites I visited on the front lines was down this trail, in the drainage of a small McKenzie River tributary called Powers Creek. I found owls that day and, on subsequent visits, determined their status as nonnesting. The habitat was, and still is, too marginal to meet their reproductive needs.

The idea of habitat as something distinct from an organism—like a stage on which they perform—is one I came to replace over the twelve years I worked with the owls. It is too reductionistic and overlooks the fact that the owl can be considered as a feature of the habitats of myriad other forms of life. When only one species is considered, the distinction between organism and habitat seems clear, but how representative of a whole ecosystem is it? And the ecosystem is, in my mind, what really matters.

In the process of putting together an assessment method for determining the habitat quality of forest groves with regard to spotted owls, I came to see the owls as one of countless shapes the forest assumes, more than as an animal that resides in a forest. In this view, if the organism is removed from the old growth, it ceases to be a spotted owl and becomes just a brown, speckled, dark-eyed, meat-eating bird.

The feature that most exemplifies this awareness for me is the owl's feathers. Unlike most raptors in this region, northern spotted owls don't migrate south in the winter. They are here in rain and snow and cold. Yet they are not exceptional thermoregulators as one would expect of a nonmigrant facing a rainy, snowy Cascade Mountain winter.

The reason for this counterintuitive disparity is the old-growth forest, which buffers temperature extremes at both ends of the spectrum by as much as twenty degrees Fahrenheit in relation to adjacent clearings. The owls wear old growth like another layer of feathers. That is, their preferred habitat provides sufficient thermal protection, so that they do not have to expend energy growing as much down as they would need if they dwelled in open country.

The deep multilayered canopy characteristic of old-growth groves also intercepts enough snow to permit the owl's preferred prey species—the northern flying squirrel—to remain active all winter long, feeding on truffle mushrooms on the open ground at the bottom of snow wells around the bases of the trees. With food and warmth (as well as many other life needs) literally covered by the old growth, the owl does not have to make a long flight to warmer climes at the onset of autumn.

Instead, the owl makes short flights in response to immediate conditions. Sun gaps on otherwise snowy January days draw them out from beneath sheltering midcanopy mistletoe umbrellas to ascend to high branches, where direct solar radiance can offer springlike warmth even during the coldest time of year. And in the heat of August, the owls are often found perched low in vine maples a few feet above a cooling creek in shady northeast-facing drainages. This behavioral thermoregulation and the incorporation of the forest itself into their meaningful physiology provide just two of many possible examples that demonstrate why efforts to reduce the spotted owl to a bird in a habitat represents extreme oversimplification.

When we are in the old growth, we are literally experiencing a meaningful feature of the body of the owl. We are within the owl's insulative layer: the protection from wind and snow we enjoy among the trees on a cold winter's day makes the forest our feathers, too.

* * *

In the evening, after my hike through the forest, I fetch dry twigs, remove two more sheets of newspaper from my pack, and carefully construct a kindling tipi. As I lean the last pieces on it, a barred owl calls from very close by. The cadence is atypical, incomplete: rather than the usual "Who cooks for you, who cooks for you all?" this one says, "Who cooks for you, cooks for you?"

Once I would have felt saddened to hear this voice in this place; visitors on field trips used to stop here to see the spotted owl pair that made this grove their home. Those spotteds have been gone for years, apparently driven away by their larger, more aggressive cousins from the east.

Barred owls arrived in the northwest only about thirty-five years ago. They crossed over from Canada, reached the Pacific Ocean and headed south down the Cascade and Coast Ranges. In appearance, barred owls differ from spotted owls mainly in their slightly larger size and in having vertically barred breast feathers instead of light spots.

During my eight years—from 1994 to 2001—as a field research assistant on the northern spotted owl demography study based at the Andrews, I saw site after site switch tenants. In a couple places, fertile hybrids, called sparred owls, turned up, but these were extremely rare.

Usually, after a short period of overlap, *Strix varia* invariably replaced *Strix occidentalis*. And I resented them for it.

There is something special—I hesitate to say "even magical"—about spotted owls. It is for them that I feel the deepest empathy, possibly in part because I spent twelve years working with them; but there's more to it than that. Maybe my strong affinity has to do with their old-growth association. This humbling, awe-inspiring ecosystem has always been my underlying motivation for learning

about, and from, the owl. The ecosystem in itself is too much—too complex, too big, too variable—to fully grasp; but when the ecosystem is crystallized into the owl, the possibility arises for humans to at least, in some small way, get our heads around it. It may be that, as Jack Ward Thomas and others have said, "Nature is not only more complex than we think, it is more complex than we can think." Our hearts are not so bounded.

The arrival of the barred owl in the old-growth forest at the apparent expense of the spotted made my heart hurt. Then my awareness changed in a single revelatory moment. I was near my house, walking along a stretch of Horse Creek Road that runs through an old-growth grove with overstory trees dated to the sixteenth century.

Above the rustling of my coat and the thudding of my footfalls on pavement, I heard the faint call of an owl. I couldn't pin down the species, but the interplay of tone and towering trees gave the impression that the forest itself was calling. I assumed a spotted owl, stopped and listened.

The forest called again.

With the voice of a barred owl.

In that instant, my opinion of barred owls underwent a profound shift, which I wrote down in a pocket notebook I had with me at the time:

> Should the old-growth ecosystem endure, even if the spotted owl gives way to the barred, the forest will retemper the barred in form and spirit to fit the mood of the trees as the spotted owl now does. It will be a softening, a quieting, a recasting for a new role in a different play of life and light and ages.
>
> Given time, the forest will do the same to inhabitant humans as befits our kind.

The barred owl calls again as I ignite the paper and blow life into my little fire. In the owl's voice is the certainty of belonging.

Human memory may recall the arrival of the species some three decades past, but to this bird, one or two generations removed from that event, *varia* has always been here, wearing this forest as *its* feathers. In the same light, so has *sapiens*. Yet if we add memory and substitute *culture* for *species*, this land has another relatively recent story of displacement to tell, a story that seems infinitely more tragic and complex, but which is really not so different, both in source and solution.

The forest lost, and needs to recover, human voices in tune with it.

That recovery is, I think, under way in everyone who comes here—researchers and reflectors alike—to learn not only about the forest but also from it, about ourselves in relation to it, as part of it, as dwellers in its shadows, seekers of its wisdom, students of its owls.

I blow again into the smoldering paper. The flames blossom, creep up into the sticks, and take hold.

I add more wood, rise from my knees and arrange my wet socks and boots to dry in the radiant heat. Then I stand toes to stones and bask as the chill of evening deepens with the shadows. The rising smoke seeks me out, stings my eyes, fills my nose with a scent humans have coaxed from wood since before we were *sapiens*.

The smoke and I dance around and around the ring while the owl calls and night falls.

A second barred owl chimes in. His deeper voice and his use of the complete cadence sets him apart from the first owl, identifiable now, in contrast, as a female. The two birds converse for a good quarter hour, then on some cue too subtle for me to detect, they fall silent. I suspect their hunting time has arrived. My stomach growls.

The dry wood soon runs out and the last coals subside. Night presses in. I crawl into my tent before the piecemeal moon has made its appearance through the trees. Since the weather is dry, I leave the rain-fly unzipped. The barred owls start up again from farther off as the final wash of twilight bleeds away. Their voices are the last thing I hear before falling asleep.

GROUND WORK

Northern Spotted Owl

Eric Forsman exchanges hoots—a *hoo . . . hoo-hoo . . . hooo*—with a northern spotted owl. It is a rainy night in the steep terrain of the Andrews Forest. The year is 1972, and the spotted owl is an obscure bird of little scientific interest. But in the decades to follow, Forsman's research on this timid, cryptic creature will totally transform forestry in the biggest timber-producing region in the country.

As a graduate student immersed in the Department of Fisheries and Wildlife's culture of hunting and fishing, it is unusual to be studying a nongame species, but Forsman pursues his task with careful, persistent observation. As the technology matures, he begins to use radio collars and telemetry to determine many locations of individual owls, and from these points he compiles a map of territories of owl pairs. He is also interested in the owl's position in food webs—what eats it and what does it eat? He carefully dissects owl pellets—the regurgitate of bones and fur of small mammals consumed by an owl—to learn what other animals the spotted owls prey upon. He studies spotted owl predators through a similar approach of patient field observations. As an outgrowth of Forsman's pioneering studies in the Andrews Forest, biologists have been tracking for decades the reproductive success of spotted owls in large, demographic study areas across their range in the Pacific coast rainforest that extends from San Francisco to the Canadian border.

Recognition of the decline of owl populations owing to loss of old-growth-forest habitat (and, to a lesser extent, to wildfire) led to its listing as an endangered species in 1990. This in turn set the stage for a judge's injunction halting timber harvest on federal forestland in the 24-million-acre range of the spotted owl. The intent of the injunction was to protect and nurture the owl's favored habitat—the old-growth forest—but it also delivered a massive hit to the timber industry. Loggers, truckers, and mill operators lost jobs as the "Forest Wars" hit their peak in the early 1990s. In a surprising twist, the endangered-species listing of the spotted owl also blew open the forest-policy window, generating new conservation strategies for hundreds of other terrestrial and aquatic species, along with whole-ecosystem strategies, such as the idea of creating a network of old-growth reserves. Scientists, including many from the Andrews Forest, played distinctive roles because the federal judge holding the injunctions relied on them for evaluation of the likelihood that alternative management options would meet policy set in the Endangered Species Act, the Clean Water Act, and other legislation.

So it was that Eric Forsman's early research into the ecology of a small, timid owl became a chief catalyst for the 1994 Northwest Forest Plan, a large-scale effort that took interagency and intergovernmental collaboration to a new level. With its bioregional, ecosystem-based conservation strategies, the Northwest Forest Plan profoundly shifted the relationship between humans and federal forestlands of the region. Nevertheless, the plan's heightened focus on habitat restoration seems to be proving inadequate for protecting the northern spotted owl, given its continued displacement by the more aggressive barred owl, an immigrant from the northeast. Thus, in the ecologically brief span of a couple of decades, the socially disruptive measures to sustain old growth in order to protect spotted owls appear to be trumped by interactions between these two owl species. What new surprises will come in the near future with changing climate, ecosystem conditions, and public perception?

The Other Side of the Clear-Cut

LAIRD CHRISTENSEN

This would be the time to turn back. I'm two-thirds of the way up a steep clear-cut, fighting for balance as the scree slips beneath my boots and clatters down toward the gravel road. From here, where the slope grows even steeper and thick with poison oak, my rented white Dodge looks like a cigarette butt. The smart thing to do would be to sidestep back down and pick a better route. Something with shade, maybe some soil, back behind the ribs of forest that line the straight edge of this clear-cut.

Instead I pick up a bleached branch to deflect the nearest threat of leaflets three, then lunge up the incline. Parry, lunge, clatter; parry, lunge, clatter. I'm making progress, but with each step I find the poison oak crowding closer. It loves these sunstruck slopes where little else will grow. No more sprays around the ankle, easy enough to avoid; here the bushes unfold to my shoulders. Now I need a stick in each hand to part the branches. I'll regret this.

Believe me, scrabbling up clear-cuts was not what I had in mind when I accepted a residency at the H. J. Andrews Experimental Forest, here in the Oregon Cascades. I was on sabbatical from a small Vermont college, working on a book about the relationships people have (or forget to have) with the places where they live. Since I grew up in Oregon, coming here might help me discover some of the ways that I'd been shaped by this landscape. That was the idea.

And what better place to rediscover Oregon forests than the Andrews? (It's always *the* Andrews, by the way, the article as inevitable as in the freeway names of Southern California.) Over the last sixty years, scientists have come to these ancient forests to study how an entire ecosystem functions, from lichens snagging nitrogen high in the canopy to the host of invertebrates building soil below. In fact, it was research conducted here that first turned the spotlight on the northern spotted owl, changing the way we use these forests.

From the moment I was invited to the Andrews, I wondered what I might add to this inspired project. Unlike previous Andrews writers—folks like Robert Michael Pyle, Robin Kimmerer, and Pattiann Rogers—I have no background in science. Despite my years spent hiking and camping in Cascadian forests, for most of my life I was oblivious to the details. Even as an adult, my study of ecology has been clumsy, occasional, and mostly self-directed.

What I *can* offer, I finally realized, is my own experience of Oregon—one that may shed some light on the competing stories that determine how we make sense of this land.

Growing up here in the 1960s, when timber was still king, I took great pride in the fact that the Laird family (on my mother's side) was among the first families who climbed down from their Conestogas, pulled out their axes, and began clearing the land. These forests played a stock role in my favorite stories as a child: wildland waiting to be made useful. By the time I returned to Oregon in the 1990s, however, after a long time away, there was a new story making the rounds. It featured people who left the crowds behind, back East, and preferred their forests upright to felled and bucked and milled.

These two stories continue to exist, side by side, sharing characters and settings despite their differences. But the hero in one story—the pioneer, the lumberjack—looks more like a villain in the other. It works the same way with symbols: a particular object can signify success in one story, tragedy in another. A clear-cut, for example. Or the northern spotted owl.

People in Oregon may share a physical landscape, but they live in symbolic landscapes that clash, sometimes violently. Through my own changes, I've seen this land as it appears in both stories. That *should* make it easier to see the symbols for what they are.

And maybe, during my time here, I can even learn to see past them.

Back in the nineties, a friend of mine, a photographer, was adjusting his tripod on the edge of a logging road, hoping to capture the light gathering on the next ridge, when a truck rumbled to a stop behind him. Two men climbed from the cab and eyed his ponytail, his Toyota, then asked what he was doing. As he explained his shot, one of the men interrupted, telling him to turn his camera to the clear-cut across the road.

"*That's* what's really beautiful," he said, and I believe he meant it.

For generations the clear-cut has been a symbol of progress, jobs, and efficiency in a region run on timber dollars. But to transplants filling up the Willamette Valley over the last few decades, the clear-cut represents all that is wrong with industrial resource-extraction.

As symbols go, the northern spotted owl is a relative newcomer. Almost overnight, once it was listed as threatened under the Endangered Species Act in 1990, this modest ball of feathers came to stand for healthy forest ecosystems. Of course, ecologists didn't call it a symbol or even a synecdoche. They preferred the term *indicator species;* but what they meant was that the survival of the spotted owl depended on preserving intact old-growth forests west of the Cascade peaks.

To folks in timber communities, fearing for their jobs, the spotted owl came to signify an unfathomable change in priorities. Those of us who traveled these back roads in the nineties couldn't miss the stickers on bumpers or at the counters of rural cafes: SAVE A LOGGER; EAT AN OWL. In fact, more timber jobs were lost to automation and policies that sent raw logs to Asia, but I guess neither fit so neatly on a bumper.

The clear-cut represents an industrial scale of human impact on the region's forests, while the spotted owl stands for our dubious ability to restrain our national economic appetite. I don't imagine many Oregonians have trouble deciding how they feel about either symbol—and the way they feel about one pretty well determines how they feel about the other.

* * *

When I was a kid, I'm not sure I even knew what clear-cuts were. I just called them "logging" and took them for granted as part of the landscape. They were my favorite playground when we visited my grandparents across the Coast Range, above the Alsea River. Head full of fantasies about those pioneer Lairds, I spent hours following elk tracks across the stripped soil and sent my hatchet spinning toward bear-sized stumps. Within a few years, that same hatchet was lopping off the live limbs I used to build overnight shelters. At nineteen I was grading lumber in a sawmill, scrawling numbers in red crayon on fresh planks of cedar. I had no problem with any kind of logging in those days.

When I returned to Oregon in my thirties, though, I saw it through new eyes. A lot had happened in the years I'd been gone, most of which I spent drifting from town to town, coast to coast, hoping to find in the next new place what I couldn't in the last. Along the way I learned that, in order to feel whole, I need something wilder than the overgrown edge of a Florida golf course. When I finally stopped for college in New England, ten years older than my classmates, my interest in wild places led me from Thoreau to Gary Snyder, Ed Abbey to Dave Foreman. I found work as a ranger and spent my summers hiking, meditating, and scribbling poems that aimed for some mystical sense of ecology. For the first time, I began to care enough about the plants around me to learn their names.

By the time a graduate fellowship led me back to Oregon in 1994, federal protection of spotted owl habitat had turned the Pacific Northwest into a battleground. Road blockades. Closing mills. Shotgunned owls. There's a reason we remember those years as the "Timber Wars."

Coming home, I was shocked by the clear-cuts that left the ridges so mangy. How had I overlooked them before? The worst I saw was on the Smith River, where the soil had sloughed off the flank of a steep clear-cut, spilling across a channel where salmon had come to spawn for thousands of years. A common poster around Eugene ran Shakespeare's words over the photo of a brutal clear-cut: "O, pardon me, thou bleeding piece of earth, that I am meek and gentle with these butchers!" Nobody needed to explain to me the symbolic power of a clear-cut.

So I drummed and shouted at protests, sent my poems to the *Earth First! Journal*, and tore flagging from the paths of future logging roads. I contributed supplies and labor to the road blockade at Warner Creek, where in 1995 activists stopped the logging of formerly protected spotted owl habitat. Most of us understood, I think, that it was not the literal owl we were fighting for but the owl as a symbol of a world left in peace.

Antagonism was so thick in the air that my time in the Oregon forests was shadowed by the possibility of violence. One afternoon, heading home from a soak in a hot spring, I watched in my mirror as a crew-cab pickup roared up behind me—the driver pissed off, I'm guessing, by the STOP CLEAR-CUTTING! sticker on the back of my truck—and began *riding* my bumper, horn blaring, before it swerved around to cut me off, braking hard and blocking the road diagonally. There was barely enough shoulder to keep me from tumbling down to the river below, but I managed to wheel around the truck just as its doors flew open, then raced to stay ahead of it, blocking its chances to pass me till finally, miles down the road, the crew gave up the chase.

After five years back in Oregon, I took my degree and then a job back East. I desperately missed the public lands of the West, however logged and grazed, but over time I learned to appreciate the woods of Vermont. True, none of those forests seem wild to a Westerner, but it's inspiring to see them re-cover 80 percent of a state that was nearly cleared a century before. I found even more promise in the middle landscape between villages and woods, where farming, recreation, and even forestry take place at a smaller scale, a slower pace. I subscribed to *Northern Woodlands*, a magazine equally interested in birding, woodstoves, ecology, sugaring, and forestry. The timber industry in Vermont is just one part of a healthy mix—not the bad guy at all.

It was only while preparing for the Andrews residency that I realized how much my time in Vermont has changed my feelings about logging. We have no clear-cuts, no spotted owls, so none of the antagonism that soured my later experience of Oregon forests.

Which is why I was so unsettled, at first, to read about the observation sites I was expected to visit at the Andrews. Oh, Lookout Creek sounded grand, down "the trail through the old growth," and I was eager to see the log decomposition site. But the third description made me squirm: "About a quarter mile from the junction the road passes through a large clear-cut on the side of a hill to the left. This is the reflections site."

It's a testament to the power of that symbol that even *imagining* being back in a clear-cut awoke the old mix of emotions: anger that felt righteous and a sadness that left me hollow in the chest, tinged with anxiety just short of fear. What would it be like to return to my old playground? To the scene of the crime?

* * *

As it turns out, not so bad—except for the poison oak.

In spite of myself, I've enjoyed this morning's hike through the clear-cut. It helps that the sun has made a surprise appearance and it finally feels like the end of May. I've stuffed my rain gear and even my shirt into my pack.

It also helps that I've been thinking about the clear-cut as a symbol, rather than letting it work on me unexamined. I'm able to set aside old feelings and see this place for what it is: about thirty acres on the south slope of Lookout Ridge, rising nine hundred feet in the quarter mile from road to ridgeline. It was logged in the 1950s and again about seven years ago.

From the moment I stepped out of my rental car, I could see that the young Douglas-fir grew best on the apron of soil above the road, with just a few adventurous saplings heading up the slope. As I cinched my pack around my waist and crunched through the gravel, the only color apparent on the hillside was the scattered canary yellow of scotch broom. I could see the fluff of deciduous trees on the ridge and a few blackened stumps along the way. Pretty desolate.

And then I noticed the flowers. Before I even left the shoulder of the road, I had pulled out *Plants of the Pacific Northwest Coast* to identify the lavender blooms at my feet. From the throat of each of three flaring petals, a yellow strip ran up the center through a fan of white. I thumbed the pages till I found a picture of an Oregon iris—and a note from the authors, Jim Pojar and Andy MacKinnon, reporting that the leaves of this flower were once braided into snares strong enough to stop an elk.

That's all it took. I spent the next two hours working my way up the clear-cut, trying to get to know each plant I found.

Who else lives in a clear-cut? There were purple peavines and tiny pink flowers that I thought must be collomia. Blue vetch sprawled around the trunks of the young Douglas-fir, whose lower limbs were draped with trailing blackberry. One common shrub I guessed must be black huckleberry, its leaves longer and more leathery than the red huckleberry I'm used to seeing. The tiny yellow trumpets among the pale bracken ferns, I decided, were monkey flowers. Halfway up the hill, I found starflowers clustered around a charred stump.

I had never imagined following a trail of wildflowers to the top of a clear-cut—but that's exactly what has brought me up the loose slope and into this swarm of poison oak.

Eventually, as I near the top, the poison oak begins to thin out, though I won't claim to have made it through unscathed. The scree that I've been climbing gives way to pale grass, featuring the fuzzy cups of mariposa lilies. At last I toss aside my sticks and step into what seems to be a natural meadow shaded by white oaks. Just beyond, between the muscular limbs of madrones and the moss-coated boulders, I see the canopy of ancient forest that fills the drainage of Lookout Creek.

I've been so intent on the flowers at my feet, and then the poison oak, that I haven't paid much attention to the vista opening behind me. Now, as I turn and lower myself to the grass, ready for lunch, I look out to the south, across the McKenzie River valley. In the distance I count thirteen clear-cuts. The largest of them, on private land above Cougar Reservoir, runs up a steep slope and along a ridge still fringed with snow. It is huge and fresh, scraped clean of life.

And just like that, the old feelings return.

* * *

A week later, on my last day at the Andrews, the clouds are just passing through, a few puffballs easing north. The sun feels so good that I leave my clothes in a pile on the gravel shore of Blue River Reservoir. Between my shallow dives and clumsy strokes, I wash away the calamine from my chest and arms, then sun myself dry against a silver log.

It's the first Friday past Memorial Day. The boaters and campers have yet to return, so it's just me and an osprey rising from the sparkles. Despite the tidy gap of a clear-cut on the ridge across the water, it's a beautiful spot.

I suppose that's a concession of sorts, making room for the human impact without ignoring how contented I feel here—drying from a dip in an artificial lake, surrounded by a forest broken up by campgrounds, roads, and yet another clear-cut. Then it hits me: almost everything I've seen during my time at the Andrews calls for a similar concession.

I had my first look at a northern spotted owl here, for example, because a biologist coaxed her from the nest with mice in order to weigh and band her. My guided tour of recent logging on public lands showed me how the forests that surround reserves of older trees are now thinned by prescription of the U.S. Forest Service, leaving behind a specific percentage of canopy and creating snags to mimic natural disturbance. For all the jackstraw jumble and decay of old-growth forest along Lookout Creek, research into the regenerative work of decomposition takes place on logs laid in orderly rows, selected portions accessible through bottomless plastic buckets.

Yes, there's an awful lot of human manipulation going on around here. Then again, seeing as how we've managed to disrupt the very climate of the planet, I suppose these disturbances don't amount to much.

History seems to tell us that human societies, once past a certain size, can't help but disturb the environment. We've seen the evidence from Sumer to Cahokia to melting ice caps. Some even suggest that there's nothing unnatural about such disturbances. After all, we're just one more species fulfilling our needs.

But other creatures live within the limits of their habitats, woven into those places by local calories and minerals, returning the nutrients in their waste and, one day, their bodies. Not us. We've done our best to insulate ourselves from natural systems. We rely on someone else, someplace else, to grow our food, mine the copper for our computers, and provide the wood that keeps us dry. We've decided that the good life requires far more than meeting our needs—and we'd just as soon remain oblivious to the consequences. Factory farms. Open-pit mines. Clear-cuts.

I think back to another clear-cut, another time. Shortly before my oral exams at the University of Oregon, I drove south past the tiny town of Wilbur, where my grandpa once ran the general store behind a dusty gas pump, then I turned east to follow the Umpqua River into the mountains. I was hoping a change of scenery might allow the preceding year's reading to settle into some useful pattern in my mind. I was also in need of context: a world big enough to remind me, after all, how insignificant those exams really were.

After a soak in a hot spring, I went searching for a camping spot, grinding up the logging roads in my little truck, slowing for each herd of elk I passed. I was after a sunny spot to burn away the chill of a Willamette Valley spring, and I eventually saw that my best option was to camp in a clear-cut—not too fresh, but not yet overgrown. It wasn't what I had in mind when I left town, but there wasn't much left to the day. Soon I had camp set up, and I spent the last hour of daylight chewing rehydrated beans and rice from the pot, looking northwest to where Eugene lay hidden beyond the foothills.

Once the sun was down, the mountains cooled quickly. I grabbed a wool hat and wrapped myself in a sleeping bag, wanting to see the full moon rise. The stars sizzled in the deepening sky, but soon I grew chilly and impatient. At last I noticed that the very edge of the ridge ahead had begun to shine. I turned to look behind me but saw no sign of the moon.

Turning back, I watched the band of light along the next ridge slowly widen as the shadow of my own ridge sunk at a nearly perceptible pace. I looked behind me: still no moon, but the sky had begun to glow above the tree line.

Soon the whole ridge before me was shining, and over the course of a magical half-hour I watched the shadow of my own ridge crawl out of the valley and creep up the hill toward me, becoming more defined the nearer it came until finally, against the exposed earth of the clear-cut, I could see the serrated edge of the shadow. I turned again and found the forested ridge in silhouette against a shining sky, but still no moon.

No, for that I had to wait until the ridge's shadow finally slid up and over my feet, my knees, my belly, and then at last I turned my face to the moonlight.

That moment, as the light washed over me, was a cleansing. Each tree below, each stump on that hillside, was likewise illuminated, and I found myself wondering: How do elk perceive such a flood of moonlight? Do they wake and glance nervously about, breath steaming in the bluish glow? What is it like for the owls, gliding above the suddenly silver boughs?

Then it struck me: as powerful as that moonrise was for me, away from town for a few days, the creatures that live here experience *every* moonrise, month after month. Those elk and all their neighbors are perfectly at home in the slow sweep of moonshadow, in the curtains of rain that rattle through, in the brightening of each dawn. That is the world they live in—and I understood, with moonlit clarity, that it's the world we live in as well.

We would *know* it is our world if we weren't so insulated by the many layers of our own cleverness. How many of us even notice the moment the moon rises above our rooftops? Oblivious to the world as it is, we close our doors and draw the curtains, slide up the needle on the thermostat, and turn our faces to the lesser worlds of our own making.

But even if we did collectively remember the world as it is, even if we owned up to the consequences of our daily lives, could we hope to limit our impact enough to strike a balance with the needs of other beings? I'm not so sure. The best we can do, at least for now, may be just to interfere in less damaging ways.

Maybe someday soon we'll reduce our demand for timber, finding more pleasure in less stuff, building our homes from cob or stone or straw, our shipping pallets from recycled milk jugs. Maybe someday soon we'll meet our remaining need for timber through the sort of artisanal forestry that flourishes in places like Wisconsin's Menominee Forest and the Vermont Family Forests. Maybe then the jobs in our timber communities will reclaim the *craft* of forestry: the judgment to know which trees to leave standing. The skill to drop and skid trees in a way that least damages their neighbors. The luxury to find the best use for each tree's unique form.

There are small choices we make every day—in our homes, at the store—that can help bring such visions to pass, and there are larger actions, too. People in the West, through ballot initiatives, can demand state laws that outlaw steep-slope clear-cutting even on private lands. We can elect federal lawmakers who will vote to stop the export of raw logs and fund the science that helps us see the forest beyond the trees. We can raise our voices, raise a sign—even raise a pickax now and then, if need be—buying time until our laws can catch up to the science. And maybe most important, as we struggle to remain hopeful, we can avoid taking for granted the very real progress we're making.

Thirty years ago a clear-cut wasn't just a symbol; it was business as usual, on public lands as well as private. Thirty years ago few people had even *heard* of the northern spotted owl, and who could have imagined that laws protecting the owl would soon change our use of public lands? How much has happened in the space of a single generation!

Lacking the hindsight of our readers in 2203, there's no way to know exactly where we're headed from here—but we'd be foolish to ignore the signs that, for now at least, we seem to be moving in the right direction.

Clear-Cut

JOAN MALOOF

Humans love the sun on a cold morning,
so do young Doug-firs,
insects in song,
and little winter wrens.

The men in the machines have given us sun,
but have taken away
liverworts, magic,
and most of the mushrooms.

A woodpecker flies by
with no place to land.
The pale stumps he sees
are dead as rocks.

Healing can be imagined,
but I will not witness it.
Only at this place do I want time to hurry,
only here the years do not go fast enough for me—
no matter how old I may be.

GROUND WORK

Forest Practices

The Andrews Forest, long known for studies of old growth, also has a long history of study about forest practices and alternative ways to manage young plantation forests. Over the years, Andrews Forest researchers and their land-manager colleagues at Willamette National Forest have launched several large-scale experiments to explore various cutting approaches and hosted many public conversations about the future of forestry in the Pacific Northwest.

The first large-scale, forest management experiment in the Andrews Forest involved clear-cutting and prescribed burning of the 250-acre Watershed 1 during the mid-1960s. The area was then planted with Douglas-fir seedlings. These were standard practices of that day, except that the size of the clear-cut was about five times larger than usual. The objective was to assess effects of these practices on vegetation, sediment production, flooding, and water yield by comparison with the adjacent "control" Watershed 2, where no logging took place. Researchers had begun their work a decade earlier by sampling the vegetation with transects of plots that cross the watershed and by measuring input and output of water and sediment. That sampling continues today. Among the many lessons from this long history of intensive inquiry, several stand out: postdisturbance vegetation has been quite vigorous, diverse, and dynamic; much of the shrub and herb vegetation is a legacy of the prelogging ecosystem; a Douglas-fir forest has gradually developed over most of the area, but even after nearly fifty years, shrub fields and rocky areas persist; and sediment yield, mainly from sources along the stream channel, continues at elevated levels. Overall, effects of the disturbance linger, emphasizing the importance of long-term study to understand the cascade of responses.

The social dimension of clear-cutting has been as dramatic as the practice itself. Starting in the 1960s, concerns about the aesthetic, civic, moral, and ecological effects of clear-cutting public lands reached a peak, culminating in legislation by Congress (the National Forest Management Act of 1976). Public perception began to shift, slowly but decisively. Still, clear-cutting persisted on federal lands, anger festered in some circles, and environmentalists continued to fight the practice.

Ecologist Jerry Franklin, in an aha moment a month after the 1980 eruption of Mount St. Helens, observed a fireweed plant sprouting from buried rootstock up through the gray volcanic ash mantling the hillsides. He understood that that small plant was a "biological legacy" of the preeruption ecosystem that would influence the course of succession thereafter. Over the decade of the 1980s, Franklin further developed this idea and proposed "New Forestry," which emphasizes *what is left* on a logged site rather than what is taken away. Retention of some live trees, standing dead trees, and downed dead wood was seen as middle ground between clear-cut and no-cut. The broad objective of New Forestry is to sustain biodiversity and productivity by incorporating features of natural ecosystems in forestry practices. The current, more nuanced version of this thinking, termed "ecological forestry," is being explored on an experimental basis, as part of the constant evolution of forestry practices driven by both new science and, especially, changing public perceptions and expectations.

Hope Tour: Three Stops

LORI ANDERSON MOSEMAN

"It may happen that we do not always want the
most beautiful form, but one of our own designing."

—SHIRLEY HIBBERD, quoted in
The Book of Topiary, by Charles H. Curtis, quoted in
"Green Animals," in *On Tact, & the Made Up World,*
by Michele Glazer

CLEAR-CUT | TEPLEY'S TOUR: STOP #1

You have to enter with a body memory of '70s' harvests
to taste this treatment as hopeful,
have to have
tallied failing seedlings in a poor reseed:
then—*Wow! They planted enough to thin.*
The more you count, the more you are accountable.
                My dogs are missing from this hike
as are hemlocks and western reds. The Incense ignored before
thrives. Cable from the clear-cut is coiled at my boot—
snake that won't burn. Public unrest
won't allow the fullest intention
to harvest again. Shall we hedge the next thirty years in a bet?
A forest's agency necessarily literal.

BLUE RIVER FACE TIMBER SALE UNIT | TEPLEY'S TOUR: STOP #2

A forester from the Philippines pronounces it ugly. Tepley
concedes:
perhaps they had wanted fire to kill fewer trees. I am happy
below the charred towers. Landform on this ridge asks for risk.
Snags and seed trees—exposed spires—say *design. Dare.*
Darkness is the awareness: it may not have been necessary—
their acting here. In this way.
We translate *squirrel*
to and from Japanese—as verb as noun as carbon budget.
Cache. Cash. We congregate. We conjugate. Computation shows
rules of conflagration are a regime tree rings already know.
Our own history not so: nothing is ever reliable 3,400 times
in a row. Not heart. Not heartwood.

OLD-GROWTH REFERENCE PLOT | TEPLEY'S TOUR: STOP #3

Further up the road within what we could have been.
Humbling. The experimental treatments not exactly
an entropy like this.

He takes the core sample and hands it to her. She breaks it,
so as to share, stores her portion in a straw she calls her magic
wand.
Earlier she had hoped to see the logging helicopter with a load.
Now, as she adds a single baby's breath blossom to her wand,
an elk appears—nearly all she had hope for
here
then gone.

Purity and Change: Reflections in an Old-Growth Forest

JOHN ELDER

I caught hold of a splintery old root and hoisted myself up onto the gray log looming beside Lookout Creek. At over five feet in diameter this fallen Douglas-fir was a tricky tree to climb, even sideways. But having caught a glimpse of an invitingly concave spot up top—just where the buttressed base had once arched out into the duff of an Oregon forest floor—I was determined to make it the viewing station for my first survey of old growth here in the western Cascades.

During the flood of 1996 the massive log on which I now sat had been scoured to the smoothness of driftwood and then wrenched perpendicular to the current. I settled in under my broad-brimmed rain hat and my slicker, watching as the ancient forest of Douglas-fir, western redcedar, and western hemlock all around me inhaled the steady rain. There seemed no limit to its lung capacity. Even when the sky cleared after a couple of hours, the thick moss draping over the undercut southern bank continued to trickle and drip for the rest of the day—a brown-green sponge through which the forest's effluent perked into the creek for its winding passage to the McKenzie River.

Lookout Creek drains the H. J. Andrews Experimental Forest, one of the world's leading sites for long-term studies of old growth. Geomorphologists run experiments here on erosion and on the dramatic surficial shifts that can occur on slopes under the influence of incessant rain. Later on in my week at the Andrews, I was lucky enough to see the celebrated, ninety-meter-high debris-flow flume in operation. Numerous scientists and field assistants (including a small contingent from Switzerland) had assembled in order to watch tons of well-watered soil, gravel, and cobbles slide down the walled ramp past cameras and electronic sensors and then spill out onto a broad concrete grid. The material sorted itself on the way down, so that the larger rocks and gravel separated into levees on either side of the muddy main current—containing the flow and thus promoting a longer course of travel before movement finally ceased. Such a dynamic was subsequently described to me by Kathy Cashman, a University of Oregon geologist, as "optimization," a process whereby nonbiological phenomena attained their maximal extension in a kind of dialogue with gravity and the laws of physics.

Another groundbreaking area of research at the Andrews is on mycorrhiza, a symbiotic association of root hairs and fungus in the upper level of the soil. An intricate network develops over many years, in which nutrients are not only transferred upward from decomposing matter in the soil but also conveyed from tree to tree. The biotic richness of old growth derives to a significant degree from the mycorrhizal web. When a fire or other natural disaster occurs, it is this belowground community that speeds recovery. The swiftness with which even forests buried under thick layers of ash after the explosion of Mount St. Helens have now recovered reflects in part the powerful functioning of fungal-root-hair networks as a biological legacy. Even more than the ancient trees themselves, these humble if generally unseen associations both sustain the character of and protect the future of old-growth communities.

Immediately adjacent to the Andrews is a 1993 clear-cut I visited on several occasions during my week. Every tree had been removed, with no apparent concern for the certainty of erosion on such a steep slope. I couldn't stop asking people I met in Oregon if there were no criminal penalties for such heedless practices. The answer, it seemed, was no. Having denuded the site, the logging company then burned off the slash and applied pesticides to hold down brush. Fifteen years later, there was little evidence of regeneration. The barren slope offered a shocking contrast to the forest flourishing just across its boundary, in the National Forest. Even more sobering, though, was the condition of the ground. Beneath thick brambles that made walking a painful challenge, the soil was as dry as powder. There was nothing there to hold moisture and foster the survival of a mycorrhizal network or any other form of biological legacy. This clear-cut illustrated, fundamentally and grievously, the eradication of old growth's millennial richness.

* * *

If subsoil research led to the most fundamental scientific discoveries at the Andrews, the understandings gained here about spotted owls have without doubt been the most politically influential. The juicy vitality of old-growth woods like these was nearly logged out of existence over the past century. Vestiges totaling about 5 percent of the original Oregon forests did ultimately receive protection, though, because of a 1990 decision to list northern spotted owls as threatened under the Endangered Species Act. A series of studies had concluded that these owls could survive only in old-growth habitat, and the resulting "threatened" listing led directly to the establishment of the Northwest Forest Plan in 1994 and to a new system of forest reserves (late successional reserves) on public lands. Thus, in addition to depending for their own continued existence on the communities of lichen-draped giants like those that rise above Lookout Creek, the spotted owls also became their main protectors.

Politically as well as ecologically, spotted owls and old growth might be said to have developed a symbiotic relationship. From the Pacific yew, western hemlock, and western redcedar growing so densely in the lower elevations of the forest to the five-hundred- to eight-hundred-year-old Douglas-firs towering above them, these woods are so dense and self-enclosed that there's no way for predatory great-horned owls to horn in on the spotted owls' niche. Old growth also abounds in the insect and fungal life that supports the spotted owls' own favored prey, like northern flying squirrels, bushy-tailed woodrats, pocket gophers, deer mice, western red-backed voles, and red tree voles. And even though private old growth has continued to disappear in Oregon, big enough patches of public old growth were secured to provide adequate habitat for this distinctive, affiliated bird. It was a triumph for conservation grounded in compelling ecological research.

It turns out, though, that the spotted owls have another problem—one that seized my attention as a visitor to the Andrews Forest from New England. At the beginning of the twentieth century, barred owls began making their way from northeastern North America across Canada to the Pacific and then extending their range southward along the coast. The second-growth woods following upon nineteenth-century timbering seemed to facilitate their migration. They are now more numerous than spotted owls in British Columbia and Washington, and their numbers are increasing in Oregon, where they have been resident since 1947. One factor in the newcomers' success has been their ability to colonize a wide range of habitats, including managed forests. They can blanket the Northwest woods, while spotted owls rely much more on the attributes of old growth—where they must also now compete for their traditional niche.

Though significantly smaller than the great horned owls, barred owls are still a bit bigger than the native owls of Oregon's old growth. A research poster in the Andrews headquarters informed me that males of the spotted owl average 582 grams and females 632 grams, while the corresponding figures for the barred owls are 637 and 801. In addition, barred owls have also proven themselves to be more aggressive in competing for territory. A research report for which Steve Ackers of the Andrews "owl team" was a principal investigator concluded that "in some areas where barred owls are particularly numerous, survey data suggest that spotted owls are gradually being displaced by barred owls and forced into marginal habitat at higher elevations. . . . The mean probability of pair occupancy decreased by approximately 5% with the presence of barred owls."

It's also now been established that spotted owls and barred owls have begun hybridizing within the old-growth forest. Whereas the characteristic hooting of the spotted owl is the four-note sequence *hoo—hoohoo—hoo,* and that of the barred owl is the nine-note *hoohoo-hoohoo—hoohoo-hoohooaaw,* the hybrid birds typically have a five-note call: *hoo—hoohoo—hooaaw.*

* * *

Upon first arriving at the Andrews forest, I had happily noted the existence of familiar plants on the floor of this remarkable and, to me, exotic landscape. White trillium was everywhere, along with yellow violets and spring beauty—in the season when those same species were flowering in the woods of Vermont. Even in the dramatically different arboreal community here, I was struck by superficial similarities. Douglas-firs—so much taller and weightier than any tree growing in Vermont, with sculpted trunks providing habitat for whole ecosystems of mosses, fungi, lichens, and insects even while still alive—nonetheless had the characteristic fir needles. The Douglas-fir is not classified as a true fir, in fact, but is placed in a separate genus of its own. Nonetheless, its short, soft, flattened needles, with occasional twists that give them a lacy look, strongly resembled the needles of our balsams. The columnar trunks of western redcedars sometimes rise to eighty meters, over three times taller than cedars in our part of New England. Nonetheless, they are clad in flattened, scrappy bark highly reminiscent of the bark of white and redcedars that thrive in wet ground and around ponds throughout Vermont's Northeast Kingdom.

Such affinities between western and eastern forests only made the presence of barred owls here more problematic. They were competing with the icons and protectors of old growth. In their hybridization with the spotted owls, too, they were blurring the distinctiveness of Oregon old growth. Above the desk in my living quarters at the Andrews was a colorful poster showing the characteristic plants, animals, trees, and birds of a "Pacific Northwest Old-Growth Forest." Right there in the foreground, looking dramatically back over its shoulder, was a spotted owl, with nary a barred owl in sight. The fact remained that, though they were intruders, barred owls were also birds with positive and highly personal associations for me.

Our family operates a sugar bush in the Green Mountain foothills of Starksboro, Vermont. In the summer we work together there to bring in the sugar wood. In the winter we walk the trails checking out our sap lines. And in early spring we gather with friends in the sugarhouse, lobbing long splits of wood into the evaporator every few minutes to maintain a high boil as we reduce forty gallons of sap to one of syrup. There are black bears denning in the upper reaches of our woods each winter, imprinting the smooth bark of beeches with semicircles of dark dots when they emerge in the spring and dig in their claws to climb after the beechnuts. Fishers stitch their weaselly tracks across the snowy slopes and squeeze under fallen trees. Coyotes launch their piercing calls across our bowl of land when night comes on. At dusk hermit thrushes sometimes sing here, too. But the soul of these woods in sugaring season is the barred owls that hoot around us in what is also their mating season. Especially when we're boiling through the night to finish up a last run while the sap is still fresh and cold, the barred owls are the midnight voice of the forest. The high, chiffy notes with which they announce themselves are inseparable from our family's excitement around the roaring fire—heat at our faces, cold air at our backs from the open door of the attached woodshed. The barred owls furnish the music in this familial ecotone, our border zone between village life and the larger life of the woods, "lovely, dark, and deep." But they mark a more alarming edge here in the Cascades, between a habitat that deepened over thousands of years and the recently fragmented and interspersed landscape that is bringing new challenges to the inhabitants of old growth.

They represent an alarming change, like the barberry, honeysuckle, and Japanese knotweed advancing into the woods of Vermont as winter comes later and ends earlier each year. Those exotic invasives take advantage of fragmentation in the forest cover of New England—spaces where stands can take hold that never could have found a foothold under more continuous cover. The warming trends that make our woods more hospitable to such species also allow for certain species of beetles, ticks, and caterpillars that had previously been frozen out by the depth of our winters. Yet another factor in the changing composition of our forests is the overpopulation of deer. They typically prefer native browse and seedlings to the invasives and thus give the newcomers even more of a competitive advantage.

* * *

The fact of the matter is that I never saw or heard an owl of any sort during my stay at the Andrews Forest in early May 2008. The winter just past had been a rather long, snowy one (as had ours in Vermont), and the researchers still were not able to get in to many of their main study sites. But that hybrid, phantom hoot (*Hoo—hoohoo—hooawwh!*) kept ringing in my ears as I learned more about this Oregon old growth and compared it to our forests back home. It made me think about how our sustainable forestry group, Vermont Family Forests, is trying to make wilderness our criterion for management and stewardship. Preserving the standing snags, protecting woodland seeps and other fragile habitats, limiting the number of roads and keeping them away from hillsides, and avoiding whole-tree harvesting and the use of synthetic pesticides are among the elements of such an approach. This means that when exotic invasives do appear, we rely upon a combination of hand and mechanical removal, selective thinning to promote a more diverse forest with a more robust canopy, and measures including both building exclosures and promoting hunting to limit deer as a factor favoring the invasives.

Even so, it is inevitable that our forests, like all forests, will change. What's more, without major pesticide programs, we'll definitely have to get used to seeing knotweed along some of our stream banks and buckthorn in our newer and lower woods. They will be a constant, and not always welcome, reminder that the natural beauty and balance we prize in our little state are altered by the same warming sky and the fragmentation of forests that affect the rest of this lovely, tilting planet. The challenge facing us is thus not only ecological but also emotional. How can people who love Oregon old growth, the returning forests of northern New England, or some other rapidly changing landscape keep from becoming downcast?

When I sought out the owl team at the Andrews to talk about the barred invaders, however, I was struck by their measured and philosophical perspective on those birds' direct competition, and interbreeding, with the indigenous spotted owls. Steve Ackers, head of that team, told me that he actually thought the competition was starting to level out, for a couple of reasons. One was that, when there's truly old-growth tree density, the barred owls are simply not quite as deft in making their way through the woods at night. As he put it, when scientists call in the spotted owls, they arrive with a nearly silent whoosh. But when the barred owls come, "it sounds as if someone just threw a football through the branches." Even the hybridization doesn't blur or weaken the spotted owls' population as dangerously as was feared at first. The hybrids tend to back-breed with the barreds much more than with the spotteds, turning, in effect, back into the barred stock. In the long run that pattern may make the local barred owls even more competitive. But it will also make them more and more like the owls with which they are competing.

* * *

There's an interesting connection (though not exactly a parallel) between the incursion of barred owls in Oregon and the belated replacement of wolves by coyotes in Vermont. We haven't had a resident wolf population in many years. As thick forests have returned to northern New England and the Adirondacks since World War II, though, we have finally regained the habitat that would support wolves. There's been a great deal of talk among wildlands proponents of reintroducing wolves here, or of taking steps to protect the ones that cross back over from Canada. But wolf scientists have increasingly argued that such policy decisions are beside the point. With the decline in hunting and trapping, populations of deer and beavers, two of the eastern wolves' main traditional prey animals, have exploded. So there's an attractive open niche for a top predator, and it's finally being filled—but not by the animal that used to occupy it.

Just as barred owls have been traveling west over the past century, coyotes have been making their way east. By the time they reach northern New England they have often interbred with Ontario gray wolves. We now are seeing larger and larger, grayer and grayer coyotes. They can still act wily and opportunistic, like coyotes in the ravines of Malibu and the foothills around Albuquerque. But they can also hunt in packs and bring down deer.

My wife, Rita, and I sometimes stay at a cabin in the Northeast Kingdom town of Craftsbury on Little Hosmer Pond. It's a shallow enough body of water that it freezes early and solid each winter, so that for several months of the year a local herd of deer can cross it going west in the morning and recross it to the east each evening. Twice in recent years we've listened to the yips and exultation as a pack of wolfy coyotes have dragged down a straggler from the herd. On each occasion, upon skiing out to the center of the pond the next morning, we found the skeleton and hide of their prey. Over the next couple of days the hide disappeared and then the bones were disarticulated. At the last there was just a pink splotch fading into the snow where nourishment had flowed out of the herd and into the life of the pack. This northern ecosystem is once more finding the predator it has needed.

The uniquely thoroughgoing and sophisticated research of scientists at the Andrews has disclosed a deeply grounded and intricately interwoven world, in which a wide array of species, including the spotted owl, the bushy-tailed woodrat, the nitrogen-fixing lichen *Lobaria*, and the truffle each play a crucial role. As long as the underlying health of the whole community remains, it may be possible for barred owls, as for coyotes in our Vermont woods, to step into the long dance without disrupting it. But the resilience of the forest itself is vulnerable, and it must be protected at all costs. The practices exemplified in the 1993 clear-cut, like the widespread use of pesticides even within a more discriminating, single-stem approach to forestry, can lead to simplification and even collapse of the underground portion of what the writer Jon Luoma has called "the hidden forest." The arrival of barred owls need not in itself spell disaster. It can simply be one among many alternations and accommodations in the ever-meandering current of energy and resources in a forest.

* * *

One advantage of taking a perspective on the forest that is grounded in biological health, yet flexible about shifts in the specific biotic community, is that it guards against too obsessive a concern for purity. Not only is avoidance of such obsession in accord with the dynamic vision of the science of ecology, but it also accords with a new, more inclusive approach within the environmental movement. The last century and a half has been a grand era in America for conservation. National parks and national forests were founded under the inspiration of figures like John Muir and George Perkins Marsh and have become essential to the culture of America and the world. In retrospect, though, there has also been a general emphasis on purity within many of the environmental achievements of this period, leading many people to sense an exclusive or privileged character in the conservation movement. The 1964 Wilderness Act with its emphasis on pristine and "untrammeled" landscapes, the Environmental Protection Agency, and the

Clean Air and Clean Water Acts all sought to prevent destruction or pollution of landscapes and biological systems. We've now come to a point, however, where such an emphasis needs to be balanced by a more inclusive, diverse, and celebratory form of advocacy.

When pursuing environmental projects, it's especially important to avoid evoking an Edenic, pre-Columbian past against which present realities have been tried and found wanting. In fact, this need has been reinforced by ecologists' own turning away from such concepts as "climax forest" to a more cyclical perspective on ecological balance. Similarly, when meeting the challenge of exotic invasives, it would especially be well to beware rhetoric that resembles that used by anti-immigrant groups! My growing discomfort with the language of environmental purity has been heightened by reading Simon Schama's *Landscape and Memory*. Schama points out that, for one seeking to identify a truly green regime in modern history, the Nazis would be a good place to start looking. They had a vision of Poland, for instance, as a huge national park roamed by native bison and wolves. There was just the detail of removing almost all the people to make that dream come true. I am certainly not calling environmentalists Nazis (though others have done so). I'm a lover of wilderness who identifies strongly with the environmental movement. But I do believe that it's time for us to trade in as much of the language of purity, exclusion, and restriction as we can manage, in order to dramatize a shift to the vocabulary of inclusion, diversity, and community.

The local-food and urban-gardening movements are two promising initiatives in this regard. Rather than being so quick to engage with nonenvironmentalists in a confrontational way ("Climb out of that SUV slowly and keep your hands where we can see them"), we should offer an invitation wherever possible ("Just taste *this*"). Farmers' markets, whether in Eugene or in Middlebury, help to support the continued existence of old-growth forests in the Cascades and second-growth woods in the Green Mountains. They celebrate the place of sustainable human communities and responsible agriculture within a larger world of healthy water, soil, plants, and animals. They stand against the assumption that there's nothing in this lovely world but commodities to be bought, sold, and consumed by economic interests already planning their exit strategies from one landscape and their extractive ventures elsewhere.

We need to do everything we can to protect native inhabitants of an ecosystem, like spotted owls, while also paying attention to the plight of rural Vermonters being pushed off the land, of city-dwellers cut off from the source of their food, and of villagers around the world subordinated to global economic systems beyond their control. When observing both ecological and social change, we need to ask what broader trends lie behind these local changes. What are the implications of such developments both for existing species and communities and for necessary alterations within the practices of our own families? Ecological and social newcomers are messengers from a turning planet. The question they pose is always the same: how we can defend the health of our world while gracefully acknowledging the inevitability of change?

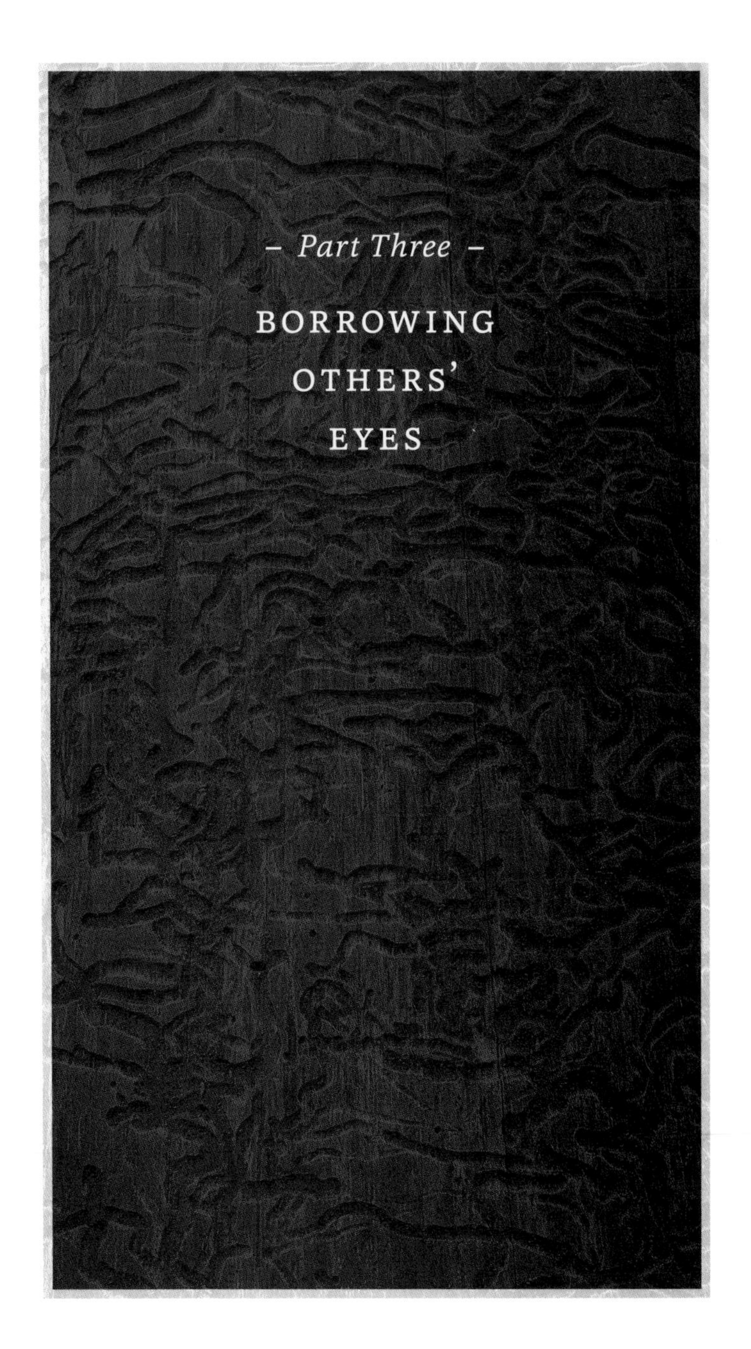

– *Part Three* –

BORROWING OTHERS' EYES

Wild Ginger

JANE HIRSHFIELD

The journey from ignorance to seeing is made by borrowing others' eyes. A friend points out the litter of disassembled fir cones under a tree in one place, and I become a person able to see them in others. Able also to see what is not, in that moment, there: the squirrel working the cone apart with fast, industrious teeth and dexterous paws, freeing its nourishing seeds.

As this spring's writer in residence in the Andrews Experimental Forest, in Oregon's western Cascades, I've been given a paper to read on the "invisible present." It points out that we perceive acutely changes that are relatively quick, but are blind to the gradual. A one-year vanishing of salmon is a recognizable crisis. If each decade a wildflower blooms a day or two earlier, we notice only if we've kept a notebook in hand.

Time is not the only governor of the invisible. Attention's direction also matters. What quiet facets of spirit have withdrawn from minds loud with overfilled hours? What creatures have vanished from memory because they once were valuable for making interesting hats? Or, conversely, because their homes were built upon without any thought for their value at all?

Pacific yew (whose bark gave us first taxol, then tamoxifen, drugs used to treat breast cancer) grows only amid the big hemlocks, cedars, and Douglas-firs of the Pacific Northwest. These trees depend in turn on a lobed, crinkled, thick-veined lichen that looks like a cabbage leaf or seaweed. Seemingly everywhere here, snagged in branches, windblown onto the ground in small curling pieces, *Lobaria* fixes nitrogen directly from air, changing it into a form plants can take up through their roots once the lichen has fallen from the canopy and been altered further by the soil's own transforming creatures. *Lobaria* takes hold only in forests eighty to a hundred years old. Yet without it, the ground's stores of nitrogen may be quickly depleted. Had there been a few more decades of ignorant lumbering, there'd be no old growth, no *Lobaria*, no yew trees performing their ancient, cronelike Noh dance in the understory's shadows, and taxol's anticarcinomic properties would have been lost. With it, the lives of friends I love. If no one is looking, no one will see.

Still, this forest's human usefulness is not the reason for the joy I feel in its presence. If the blizzards and brunt of almost sixty years have taught me anything that can be spoken, it is to want what the poet Robinson Jeffers described: "I have fallen in love outward." Unremitting self-concern suffocates. As with the body's own breathing, the gestures of mind and heart that sustain are the ones that include, exchange, reach outward with hands held open. In this place, I feel a life whose center is the whole, the commons of air, water, light, soil, animals, insects, incline, stone. An old-growth forest—incomprehensible, multistoried, particular—is not ours to manage or mine. We are its. What relief that recognition brings.

In the physics of ancient Greece, Love was said to hold the world in place against Hate's entropic dismantling. Gravity, I've been told, is a "black box" concept, still not fully understood. "The strong force," "the weak force," "strange attractors" . . . why not call these quantum connections love? It's as accurate as any other name for what we know of the world.

Each individual scale of a Douglas-fir cone wears the shape my friend showed me: the rear half of a fast-moving mouse whose front end has burrowed into the pale brown bract. Two pointed, racing hind feet; a tail so straight it extends past the end of the scale—there is no way to tell if the mouse is returning to a long-familiar refuge, seeking the cone's hidden seed, or fleeing this troubled invisible present into a future equally unknowable and unknown. All I can see is a story of change and connection: one being-form becoming, in front of my eyes, another. Mouse racing into the fir-cone pattern that makes it.

Nearby, wild ginger is growing. Good for tea, my friend says.

This Day, Tomorrow, and the Next

PATTIANN ROGERS

When the blind and the deaf walk
together into the forest, one of them
understands the blackness of light
on a clear day. The other understands
the deep reach of stillness in a riot of green.

Both alike feel on their faces
the floating threads and tatters
of occasional sun passing through
the canopy of overlapping branches,
close thatch of needles, uneven roof
of broad leaves. And both can name
the fragrances of sweet sap and damp
soil, sodden cones, rain-filled mosses.

But neither encounters the burrow
of the fungus beetle leaving her eggs
in the dank of a fallen fir. Neither
is aware of the yellow of the jewelweed
to come. Neither is aware of the taste
of the salmonberry to be. Neither imagines
the spirit-deer made of thicket shadows,
the deer known only when imagined.

Within their inevitable errors
each regrets, each beholds.
Both put their hands the same

into the snow waters of the creek,
the flow pushing equally against
the pressure of their place.
But only one tilts toward the single
twitch-sluff of ground leaf where
the red newt slides. And only
one of them finds and lifts the red
newt from its rust-red leaf.

Each can hold a river-smooth
rock, feel the circle-leading allure
of its edges, remember by finger
and palm the shape and heft
of before the beginning. Within
their frailties, each asserts, each fears.

And for a brief moment either
of them might conceive and come
to love that which exists solely
as the possibility of radiant
green fern-leaf fronds spreading
over the forest floor, yellow-green
and black-green fir and cedar,
hemlocks filled with hanging moss
shags, the possibility of a ruffled
spill of lichens, the rip of a steel
blue creek, the chip *zeet* of dipper,
the slow swing of autumn fog
up the hillside, conceive and love
that possibility alone which attends
steadily without ears and watches
forever without eyes.

Portrait

Parsing My Wife as Lookout Creek

ANDREW C. GOTTLIEB

My wife sits, wipes, stands, zips, forgets to flush.
Rushing,
the river's every agenda. We pull at our clothing,
all day, humans, us,
all of us.
Try not to touch it.

I stand at the mirror, tuck a tail, a tag, tug a collar, flinch.
What face is that?
Dry
outside, there are pines pushing against every reflecting sky
in their own grim time.

My mother, tough one, British stiff. *Sit up straight. Excuse*
you. That's a dessert spoon.
Butler's fool,
ambassador for a childhood of rules. One tough one.

Language gets us in its grip with its little links and latches,
clasps, clamps,
padlocks,
and we're lost: grappling.

Close your mouth when you chew.
In these river days,

what floats for me to find is the tissue, wet, a red filmy swirl
the symptom of a drifting of cells
alluvial shift
in a body I know.

Do you imagine first the conifer leaves?
Or the buried threadlike roots
deeply reaching for food?
Plunging to touch the hidden skin
of the river.
Dawn's lazy diffusion of hues lights the children's
confusion, their breakfast food,
flow
of this river that spews
stripped trunks, a shoe, crescent crust of dead everything,
the ongoing plunge of innard and corpse.

Even my stepdaughter laughs, who for now laughs last,
least.
There's nothing funny about PMS: period.
My wife,
sure, she blushes, but it's love like the cat's torn mouse,
the breast-split wren,
the rejected owl pellet,
her kind of love,

the river's necessary way of sharing of what she's composed,
unburdened by grammars, maps, latitudes, rules,
banks.
I am wading
the lava rock and free-stone bed,

the old-growth bole
wedged
and lecturing only by collecting

every drifting thing that the muscle spits up, aggregate of flow,
   motion of bundling,
clustered abundance of the rushing's best refuse.
                                                            I steady my step,
pocket a bottle, sift the river with my fingers, sink
   into its stunning flood,
                                        touch her every part.

On Assignment in the H. J. Andrews, the Poet Thinks of Her Ovaries

MAYA JEWELL ZELLER

From the base of a Douglas-fir, the forest
management guru is saying important things
like *biodiversity* and *water quality*,
but I can't help listening to that creek, the birds,
buzzing insects, the silence the trees make,
light filtered by their needles. Or the silence is you,
and I am the trees, making you, spinning you out
into this perfect golden air. Here I am thinking
of you, ten years old, how you'll play
in the stream while shadows weave
through your toes, minnows instruct
your next move. Now you are unformed,
an egg, the sperm that will start you still growing
back in Spokane, the man who will be your father
with his quiet pulse. I keep imagining you,
the way you are the rays passing through
hemlock, the way you warm it,
coach it to accept the moss,
make your way to its belly, the way it lives
whether or not you ever enter this world.

Piles of Pale Green

JOSEPH BRUCHAC

All day yesterday and the day before, warm spring wind rippled through this fold between the mountains that holds the Andrews Forest buildings like small blocks of wood in a giant's hand. That wind brought down a dry rain of small branches bearing wispy banners of epiphytic lichens. They nearly covered the road in places like blankets textured with green and gray, woven by the dark threads of the twigs. Pieces of *Lobaria* that look like poorly sewn-on patches fallen from an old man's tattered coat.

They still have life, those old-men's beards—like the Spanish moss of the bayous of the southeast. When the same wind that brings life to my lungs touches them, they quiver, little flags still representing their nation, ones that do not like to let go, that cling to the branches, the trunks, anything that gives them purchase. Here in this forest, life attaches itself to life. Even the gate that has been swung back and latched open is plastered with green moss, soft and spongy and drier to the touch than the rusting metal hidden beneath it.

I gathered up some of those fallen gifts from the high branches, knowing what they are, yet my eyes and my brain are still fooled every now and then into seeing them as squirrel tails. A pile of them rests now on the table next to me, like trophies taken after a day of hunting for meat.

Was there a time that I remember when squirrel tails would be hung from the radio antennas of cars? I know that during my years in grad school, when I hunted squirrels I sometimes saved their tails after skinning their warm bodies, feeling the slickness of firm red muscle beneath my fingertips, slicking out their guts and returning them to a hole dug in the leaves, then taking the hindquarters to our two-room Quonset-hut apartment, where my wife, Carol, would fry them. We were poor then by the standards of many, and our meals often centered on what I could gather from the woods and fields or take with my grandfather's old pump-action twelve-gauge. Cattails, dandelion greens, and milkweed in the spring. Pheasant, partridge, squirrel, the berries that came in late spring and summer and fall. Strawberries, the gift of the Little People. Raspberries as red as blood, blackberries hanging in dark clusters on heavy vines, blueberry bushes so laden with fruit we'd fill buckets with them as we gathered on the top of Turkey Hill. We ate them and gave thanks for their lives sustaining our lives.

Cycles. I think of cycles as I pick up one of those squirrel tails of lichen wrapped around the brown bone of a Douglas-fir twig. Car wheels run over those fallen bits of the high forest, human feet kick them to the side to clear the pavement. But they are the flags of a nation that is undefiled by being broken, being torn, being crushed into the soil. They fall bearing their gifts of nutrients, nets that hold the earth, the roots, the healing rain.

I press a handful of them against my face, inhale the clean, dry, almost animal scent. A part of me wants to keep them, to never let them go. But I know that is a foolish wish. For they symbolize nothing except themselves, and their use is not to be hoarded away as human possessions, as tangible metaphors. A small laugh escapes my throat as I think that, because my eyes have drifted to another pile of pale green on that same table—the currency I took from my pocket and dropped there last night. A twenty and four ones. The legendary amount supposedly paid by the Dutch to purchase Manhattan on May 24 of 1626. Sixty guilders, actually, an amount equated to that fabled twenty-four simoleons. Money for land, an exchange that has never been truly understood—and why should it make sense?—by the native people of this hemisphere, where the earth is often seen and experienced as a relative, a sustaining parent, and not something to be cut up and consumed like a piece of meat.

I pick up the twenty. My least-favorite bill, for it bears on its face the face of Andrew Jackson, a man whose life and fortunes were saved by his Muskogee allies. Men who, with their families and those of the other tribal nations of the American South, were forced from their homes and sent toward the setting sun on paths watered by the tears, sweat, and blood of many generations. My Cherokee friends still refer to Jackson as the devil. It was Jackson who, slightly more than any other president, saw Indians as an impediment to be removed.

Then I take up the ones, the most common of all bills. And on their faces is the visage of the Father of Our Country. The man whose name was placed on the nation's capital. And when you say the word *Washington*, you are doing more than just referencing the first president or even that city. You are talking about policy and politics, about the way things work, or do not work, in America.

Rex Jim, a Navajo friend, once asked me if I knew the right way to hand someone a dollar bill. "Like this," he said, holding it out green-side up. "The green and the eagle are on this side. You can always trust the eagle and the green." He flipped the one over. "But you can never trust the man, trust Washington."

I weigh that small pile of bills in one hand and the little bundle of twigs, lichen, and mosses in the other. Which one can I hold on to, truly hold? Which one can I carry? And which, when I put it down, will carry me?

Design

JERRY MARTIEN

For Charles Goodrich and Clemens Starck

Three old poets on a wooden footbridge. Admiring its pole construction. Its craftily fitted joints. Pieces shaped and cut and assembled to span the creek rushing beneath us.

The trail winds among the massive trunks of old-growth Douglas-fir. Cedar. Yew. Standing. Fallen. Decomposing into duff. Spared the clear-cuts of the surrounding mountains, the millions of years go on. Refugia, Charles says. Places where things go on.

He bends down and picks up a leaf of prehistoric lettuce. Hands me a piece. *Lobaria*, he says. Ancient lichen, pockets and ridges like the landscape of another world. Now I see it in the canopy above us, scattered on the forest floor. Green and gray. Thriving and dying. A fungus and an alga hooked up with a bacterium. It can photosynthesize, fix nitrogen from the air, reproduce from spores or broken pieces of itself. Its decay feeds the forest with nitrogen. So it goes on. He hands a piece to Clem.

Who stops beside a yew. A dense centuries-old branch his hand can reach around. The English longbow. Battle of Agincourt, he says. Then: Henry V? Laughs at the French general whining about the rules of war. Clem admires a good tool.

Then devil's club, Charles says. Points to spiny stalks thrusting up through the forest floor. With wrapped burlap for a handhold, company goons beat the Wobblies when they ran them out of Everett and other mill towns. I touch one, its spines sharp as needles. Ingenious, I say. A moment of silence for the IWW.

At a wide spot in the trail three old poets looking up at a spider's web.
Strung between yews on either side, an artful airy construct, its author
in the middle of it. Pollen from the firs has dusted the spider and every
strut and strand with gold. All that beauty. Useless now that it's visible.
Poor guy, we all commiserate.
He'll never get dinner with that.

Listening to Water

ROBIN WALL KIMMERER

I feel my number acutely. Two of two hundred, the second writer to be summoned to this place. The second voice in a chronicle intended to stretch out for two hundred years. Two hundred years is young for the trees whose tops this morning are hung with mist. It's an eyeblink of time for the river that I hear through my open window and nothing at all for the rocks. The rocks and the river and these very same trees will likely be here for the two-hundredth writer, if we take good care. As for me, and that chipmunk, and the cloud of gnats milling in a shaft of sunlight, we will have moved on.

Long-term ecological reflection. It's tempting to let your thoughts run out to the future and back into the past, to reach for the stories that live there. Time as objective reality has never made much sense to me. I don't wear a watch, and I slide between time and timelessness with ease. How can the minutes and years, millennia and nanoseconds, devices of our own creation, mean the same thing to gnats and to cedars? The particular moment and the collective years seem to bear little relation to one another. It's only what happens that matters. If there is meaning in the past and in the imagined future it is captured in the present. To be part of long-term ecological reflection is, then, to be quite literally a mirror on a moment and then let time take care of itself.

When you have all the time in the world, you can spend it, not on going somewhere, but on being where you are. And observing the life of raindrops. In my bright yellow slicker, I walked through the forest and down to the stream. Listening.

This Oregon rain, at the start of winter, falls steadily in sheets of gray. Falling unimpeded it makes a gentle hiss. You'd think that

rain falls equally over the land, but it doesn't. The rhythm and the tempo change markedly from place to place. Standing in a tangle of salal and Oregon grape, I listen as the rain strikes a *ratatatat* on the hard, shiny leaves, the snare drum of sclerophylls. Rhododendron leaves, broad and flat, receive the rain with a smack that makes the leaf bounce and rebound, dancing in the rain. Beneath this massive hemlock, the drops are fewer, and the craggy trunk knows rain as dribbles down its furrows. On bare soil the rain splats on the clay, while fir needles swallow it up with an audible gulp.

In contrast, the fall of rain on moss is nearly silent. I kneel among the mosses, sinking into their softness to watch and to listen. Rain falls all around me, but all I can hear is the splat on my raincoat. The drops are so quick that my eye is always chasing, but not catching, their arrival. At last, by narrowing my gaze to just one single frond and staring, I see it. The impact bows the shoot downward, but the drop itself vanishes. It is soundless. There is no drip or splash, but I can see the front of water move, darkening the stem as it is drunk in and silently dissipated among the tiny shingled leaves.

Most other places I know, water is a discrete entity. It is hemmed in by well-defined boundaries, lake shores, stream banks, or the great rocky coastline. You can stand at its edge and say "this is water" and "this is land." Those fish and those tadpoles are of the water realm; these trees, these mosses, and these four-leggeds are creatures of the land. But here in these misty forests those edges seem to blur, with rain so fine and constant as to be indistinguishable from air. Cedars wrapped with cloud so dense that only their outline forms emerge. Water doesn't seem to make a clear distinction between gaseous phase and liquid. The air merely touches a leaf or a tendril of my hair and suddenly a drop appears.

Even the river, Lookout Creek, doesn't respect clear boundaries. It tumbles and slides like an ordinary creek down its main channel, where a dipper rides between pools. But Fred Swanson, a hydrologist here at the Andrews, has told me stories of another stream, an invisible shadow of Lookout Creek, the hyporheic flow. This is the water that moves under the stream, in cobble beds and old sandbars. It edges up the toeslope to the forest, a wide unseen river that flows beneath the eddies and the splash. A deep invisible river, known to roots and rocks, the water and the land, intimate beyond our knowing. It is hyporheic flow that I'm listening for.

I wander the paths along the shore just looking. I lean up against an old cedar with my back nestled in its curves and try to imagine the currents below. But all I sense is water dripping down my neck. Every branch is weighted down with curtains of *Isothecium*, and droplets hang from the tangled ends, just as they hang from my hair. When I bend my head over, I can see them both. But the droplets on *Isothecium* are far bigger than the drops on my bangs. In fact the drops of moss water seem larger than any I know, and they hang, swelling and pregnant with gravity, far longer than the drops on me or on twigs or bark. They dangle and rotate, reflecting the entire forest and a woman in a bright yellow slicker.

Can I trust what I think I'm seeing? I wish I had a set of calipers, so that I could measure the drops of moss water and see if they really are bigger. But surely all drops are created equal? I don't know, so I take refuge in the play of scientists, spinning out hypotheses. Perhaps the high humidity around moss makes the drops last longer? Maybe in their residence among mosses, they absorb some property that increases the surface tension of the drop, making it stronger against the pull of gravity? Perhaps it's just illusion, like how the full moon looks so much bigger at the horizon. The diminutive scale of the moss leaves makes the drops appear larger? Maybe they want to show off their sparkles just a little longer?

After hours in the penetrating rain, I am suddenly cold and damp. Where are the dry places, I wonder? Surely there are niches here and there where the rain does not reach. I poke my head into an undercut bank by the stream, but its back wall runs with rivulets. No shelter there, nor in the hollow of a treefall, where I hoped the upturned roots would slow the rain. A spiderweb hangs between two dangling roots. Even this is filled, a silken hammock cradling a spoonful of water. My hopes rise where the vine maples are bent low to form a moss-draped dome. I push aside the *Isothecium* curtain and stoop to enter the tiny dark room, roofed with layers of moss. It's quiet and windless, just big enough for one. The light comes through the moss-woven roof like pinprick stars, along with the drips.

As I walk back to the trail, a giant log blocks the way. It has fallen from the toeslope out into the river, where its branches drag in the rising current. Its top rests on the opposite shore. Going under looks easier than going over, so I drop to hands and knees. And here I find my dry place. The ground mosses are brown and dry, the soil still soft and powdery. The log makes a roof overhead more than a meter wide in the wedge-shaped space where the slope falls away to the stream. I can stretch out my legs, the slope angle perfectly accommodating the length of my back. I let my head rest in a dry nest of *Hylocomium* and sigh in contentment. My breath forms a cloud above me, up where brown tufts of moss still cling to the furrowed bark, embroidered with spiderwebs and wisps of lichen that haven't seen the sun since this tree became a log. This log, inches above my face, weighs many tons. All that keeps it from seeking its natural angle of repose upon my chest is a hinge of fractured wood at the stump and cracked branches propped on the other side of the stream. It could loose those bonds at any moment. And one day it will. But given the fast tempo of raindrops and the slow tempo of treefalls, I feel safe in the moment. The pace of my resting and the pace of its falling run on different clocks, so I stretch out, close my eyes, and listen to the rain.

The cushiony moss keeps me warm and dry, and I roll over on my elbow to look out upon the wet world. The drops fall heavily on a patch of *Mnium insigne*, right at eye level. This moss stands upright, nearly two inches tall. The leaves are broad and rounded, like a fig tree in miniature. One leaf among the many draws my eye by its long tapered tip, so unlike the rounded edges of the others. As I lean in closer, my head lines up with the drip line of the log, but no matter. The threadlike tip of the leaf is moving, animated in a most unplantlike fashion. The thread seems firmly anchored to the apex of the moss leaf, an extension of its pellucid green. But the tip is circling, waving in the air as if it is searching for something. Its motion reminds me of the way inchworms rise up on their hind sucker feet and wave their long bodies about until they encounter the adjacent twig, to which they then attach their forelegs, release the back and arch across the gulf of empty space. But this is no many-legged caterpillar; it is a shiny green filament, a moss thread, lit from within like a fiber-optic element. As I watch, the wandering thread touches upon a leaf just millimeters away. It seems to tap several times at the new leaf and then, as if reassured, it stretches itself out across the gap. It holds like a taut green cable, more than doubling its initial length. For just a moment, the two mosses are bridged by the shining green thread, and then green light flows like a river across the bridge and vanishes, lost in the greenness of the moss. Is that not grace, to see an animal made of green light and water, a mere thread of a being who like me has gone walking in the rain?

Down by the river, I stand and listen. The sound of individual raindrops is lost in the foaming white rush and smooth glide over rock. If you didn't know better, you might not recognize raindrops and rivers as kin, so different are the particular and the collective. I lean over a still pool, reach in my hand and let the drops fall from my fingers, just to be sure; and indeed there are drops in that river, and both the hanging drop and the surface of the river reflect back to me the forest.

Between the forest and the stream lies a gravel bar, a jumble of rocks swept down from high mountains in a river-changing flood last decade. Willows and alders, brambles, and moss have taken hold there, but this too shall pass, says the river.

Alder leaves lie fallen on the gravel, their drying edges upturned to form leafy cups. Rainwater has pooled in several leaves and it is stained red brown like tea with tannins leached from the leaf. Strands of lichen lie scattered among them where the wind has torn them free. Suddenly I see the experiment I need to do to test my hypothesis; the materials are laid before me. I find two strands of lichen, equal in size and length, and blot them on my flannel shirt inside my rain coat. One strand I place in the leafcup of red alder tea, the other I soak in a pool of pure rainwater. Slowly I lift them both up, side by side before my face and watch the droplets form at the ends of the moss strands. Sure enough they are different. The plain water forms small, frequent drops that seem in a hurry to let go. But, the droplets steeped in alder water grow large, heavy, and hang for a long moment before gravity pulls them away. I feel the grin spreading over my face. There are different kinds of drops, depending on the relationship between the water and the plant. If tannin-rich alder water increases the size of the drops, might not water seeping through a long curtain of moss also pick up tannins, making the big strong drops I thought I was seeing?

Where new gravel meets old shore, a still pond has formed beneath the overhanging trees. The shallow basin is cut off from the main channel, but it fills from the rise of hyporheic flow beneath it. Summer's daisies look surprised to be submerged two feet deep now that the rains have come. In summer, this pool was a flowery swale, now a sunken meadow that tells of the river's transition from low, braided channel to the full banks of winter. It is a different river in August than in October. You'd have to stand here a long time to know them both. And even longer to know the river that was before the coming of the gravel bar, and the river that will be after it leaves.

Perhaps in just this moment, we cannot know the river. But what about the drops? I stand for a long time, quietly, by the still backwater pool and listen. It is a mirror for the falling rain and is textured all over by its fine and steady fall. I strain to hear only that sound among the many, and find that I can. It arrives with a high sprickley sound, a *shurrring* so light that it only blurs the glassy surface, but does not disrupt the reflection. The pool is overhung with branches of vine maple reaching from the shore, a low spray of hemlock, and, from the gravel bar, alder stems incline over the edge. Water falls from each of these into the pool, each to its own rhythm. The hemlock makes a rapid pulse. Water collects on every needle but travels to the branch tips before falling, running to the drip line, where it releases in a steady *pit, pit, pit, pit, pit,* drawing a dotted line in the water below.

Maple stems shed their water much differently. The drips from maple are big and heavy. I watch them form and then plummet to the surface of the pool. They hit with such force that the drop makes a deep and hollow sound. *Bloink.* The rebound causes the water to jump from the surface, so it looks as if it were erupting from below. There are sporadic *bloinks* beneath the maples. Why is this drop so different from the hemlock drips? I step in close to watch the way that water moves on maple. The drops don't form just anywhere along the stem. They arise mostly where past years' bud scars have formed a tiny ridge. The rainwater sheets over the smooth green bark and gets dammed up behind the wall of the bud scars. It swells and gathers until it tops the little dam and spills over, tumbling in a massive drop to the water below. *Bloink.*

Sshhhhh from rain, *pitpitpit* from hemlock, *bloink* from maple, and lastly *popp* of falling alder water. These drops come more rarely, a slow music. It takes time for fine rain to traverse the rough surface of an alder leaf. Its drop is not as big as a maple drop, not enough to splash, but its *popp* ripples the surface and sends out concentric rings. I close my eyes and listen to all the voices in the rain.

The reflecting surface of the pool is textured with their signatures, each one different in pace and resonance. Every drip it seems is changed by its relationship with life, whether it encounters moss or maple or fir bark or my hair. And we think of it as simply rain, as if it were one thing, as if we understood it. I think that moss knows rain better than we do, and so do maples. We should be listening. Maybe there is no such thing as rain; there are only raindrops, each with its own story.

Listening to rain, time disappears. If time is measured by the period between events, alder drip time is different from maple drop. Isn't it likely that the surface of this forest is as textured with different kinds of time, as the surface of the pool is textured by different kinds of rain? Fir needles fall with the high-frequency hiss of rain, branches fall with the *bloink* of big drops, and trees with a thunderous thud, so rarely. Unless you measure time like a river, which catches trees many times in its life. And we think of it as simply time, as if it were one thing, as if we understood it. Maybe there is no such thing as time; there are only moments, each with its own story.

Every drop has a life of its own, a story colored by its meetings, and we can no more see the drops in the river than we can see the river which shifts from season to season, year to year, so that we scarcely know it. And so, you two hundred years' worth of writers, how blessed we are to have a chance to go beyond ourselves, to listen to the stories of this place, to give voice to rain. Our pages may stretch like a mysterious being reaching across the gulf, connecting us. We need each other, all the voices, yours and mine, hemlocks', sword ferns', gnats', ravens', and green-thread animals', to tell the stories, to hand them down so that we might come to know the river and together ride its currents to safety.

GROUND WORK

Water

In winter, the entire Andrews Forest world is aquatic. Rain falls incessantly—up to one hundred inches per year on average. Clouds swirl through the treetops; moisture gathers in the foliage and drips to the ground (hydrologists call it "throughfall") or flows in rivulets down the furrowed tree bark (called "stemflow"). Aquatic bugs inhabit thick, soggy mats of moss draping tree limbs high above the stream. Amphibians can roam the saturated everywhere. People experience the same, pervasive dampness. If you walk hard wearing rain gear, you're drenched with sweat from the inside, and a trickle bisects your shoulder blades. Every seam sucks in water from outside. Water is as characteristic of the landscape as the forest is. Rainforest could as rightly be called waterforest. Water is habitat, essential for life, a source of sound, and an agent of disturbance in either of its extremes, of scarcity or abundance.

Hydrologists like to ask, "Where does the water go when it rains?" Understanding the flow of the abundant precipitation that falls as rain and snow would seem simple enough, but it turns out to be quite complicated. Researchers at Andrews Forest, like their counterparts at many other places, began pursuing hydrology studies in the 1950s in three adjacent watersheds of approximately one hundred hectares (two hundred fifty acres). They treated them as "black boxes" and simply measured water input with a precipitation gauge and output at a stream-gauging station. After five to ten years of sampling, one of these experimental watersheds was designated as an untreated "control," and others were subjected to various combinations of logging practices (euphemistically termed "treatments") and road construction. Streamflow measurements before and after these management disturbances tracked variation in floods, low flows, and total water yield in response to, first, the removal and then the regrowth of vegetation in relation to the control watershed. Then, researchers began to disentangle the intricacies of the many ways and places that water is stored within a watershed and the various processes that move water from one location to another. When rain or snow falls into the forest, some is intercepted by the canopy and evaporates or sublimates back to the atmosphere; and some collects in the foliage, accumulating in large droplets that fall to the forest floor as throughfall along with the smaller raindrops. Water entering the soil makes a slow passage to streams and departs the watershed. Some of the soil water is taken up by trees and pumped upward to be expelled from stomata (little pores on the underside of needles) back into the atmosphere via the process of transpiration. Researchers employ many types of instruments, long-term record keeping, and methods of isotope chemistry to measure these and other stores and fluxes.

Studies of watershed hydrology that began in the 1950s emphasized forestry hydrology—the effects of forest practices on streamflow. In recent decades, research has expanded to include basic headwater ecosystem studies concerned with how vegetation and climate affect the quantity and quality of water runoff, the carbon sequestration on site, and other "ecological services." Forests play a major role in the water cycle. Approximately 45 percent of annual precipitation is returned to the atmosphere via evaporation of water that landed on foliage and limbs and by the piping of water up through the stem. Hydrologists try to sort out these mechanisms by direct measurements of water flow up trees and indirectly by observation of changes in streamflow when trees are removed from experimental watersheds. Other studies delve into previously unrecognized water flow-paths, such as the river below the river—the hyporheic system of subsurface water movement within sediment in the valley floor.

For the Lobaria, Usnea, *Witch's Hair, Map Lichen, Ground Lichen, Shield Lichen*

JANE HIRSHFIELD

Back then, what did I know?
The names of subway lines, buses.
How long it took to walk twenty blocks.

Uptown and downtown.
Not north, not south, not you.

When I saw you, later, seaweed reefed in the air,
you were gray-green, incomprehensible, old.
What you clung to, hung from: old.
Trees looking half dead, stones.

Marriage of fungi and algae,
chemists of air,
Changers of nitrogen-unusable into nitrogen-usable.

Like those nameless ones
who kept painting, shaping, engraving
unseen, unread, unremembered.
Not caring if they were no good, if they were past it.

Rock wools, water fans, earth scale, mouse ears, dust,
ash-of-the-woods.
Transformers unvalued, uncounted.
Cell by cell, word by word, making a world they could live in.

The Owl, Spotted

ALISON HAWTHORNE DEMING

Steve Ackers and I clamber over vine maple and Oregon grape, a tangled mess of scrub that covers Hardy Ridge high over Cougar Reservoir in Oregon's western Cascades. This terrain is better suited to flying squirrels and red-backed voles than to a mildly arthritic, bipedal primate. But here I am on a sun-drenched morning in mid-May, hiking with the head of the northern spotted owl research team from the H. J. Andrews Experimental Forest and filled with unaccountable joy. Last night Steve was orienteering toward an owl that was calling from a mile away. He set a compass point and hiked into the dark forest toward the call, but never found the bird. He's been working on the owl study for seven years, on wildlife fieldwork for twenty-one. Today we're looking for a spotted owl that has been in the study for twelve years, one habituated to the visits of field scientists.

Extensive study of this species had been conducted for at least a decade prior to its 1990 designation as threatened under the Endangered Species Act. The northern spotted owl is perhaps the most studied bird in the world, inspiring unprecedented collaboration among scientists, federal and state agencies, universities, and landowners.

We break into an opening shaded by a small stand of Douglas-fir—trees not super-old, as we've seen along the McKenzie River Trail, where there are giants six hundred years old, but stately elders nonetheless. The ground is dappled with light, the air cool and damp. The hillside slopes steeply below. Ahead of me Steve hoots the four-note location call: *hooh-hoohoo-hooooh*. The last syllable descends with a slight warble. No response. Then he turns and a quiet smile opens on his face. He has the bright and easy look of a man who knows how lucky he is to love his work. He points over my left shoulder.

Silent, she's perched on a small understory branch twenty feet up. She's watching us, waiting for us to notice her. She knows the contract. She will give us data, we will give her mice. After three decades of research on the northern spotted owl, scientists have gained a wealth of understanding about this creature's life history. Each spring the field crew checks nesting pairs for their reproductive status and bands fledglings to include them in future surveys. The data gathered led in 1994 to the comprehensive Northwest Forest Plan, which decreased the rate of logging and altered how it is done, giving the owls and their entire ecosystem a better chance at survival. But data cannot compare to the experience of that deep well of attention, quiet, and presence that is the owl. She has a spotted breast; a long, barred tail; and tawny facial disks with brown semicircles fringing her face and back-to-back white parentheses framing her eyes. These markings give the impression that her eyes are the size of her head. The blackness of her pupils is so pure they look like portals into the universe.

When Steve takes the first mouse out of his aerated Tupperware container, lifting it by its tail and placing it on a log, the owl drops, silent as air, down through the branches and closes her talons. She lofts back up to the branch and scans around. She may be looking to see if a goshawk is near. Whatever constitutes a threat to her does not include us. How rare it is to have more than a fleeting glimpse of a creature in the wild. Still clutching the mouse, she burps up a pellet that plops to the ground, gives us a nonchalant look, then gulps down her meal.

"You want to see the parachute drop?" Steve asks with a grin. He places a second mouse on the log, and she billows out her wings, buoying herself down to us. It takes a moment to understand why her flight catches me each time by surprise. No riffle, no flutter of resistance through the feathers, she's evolved for this easy drop onto her prey. The spotted owl is a sit-and-wait hunter, unlike the goshawk, which will tear through the woods in pursuit. The fringed edge to her wing reduces noise and increases drag, making this strategy a good match of form with function.

Steve collects the pellet and we poke apart the slimy gray glob of indigestible fur and bones from the past day. The bones are very delicate, still shiny with the life that left them, some nearly two inches long.

"Maybe a wood rat," Steve says. Through binoculars he can see the owl's identification band. Last year a male was keeping this female company, a two-year-old from King Creek. This year, so far, she appears to be alone. The owl team's last visit to this site was one month ago.

"How about the side grab?" Steve asks. He might be a dad boasting about the agility of his soccer-playing daughter. He isn't making the owl perform for our enjoyment. These flight skills are as natural to her as stepping over a crack in a sidewalk is for us. The mouse is barely out of his hand, scurrying in confusion on the tree trunk that rises beside me, when the owl swoops onto it, talons leading, and picks it off. It happens so fast that she's flying away by the time I realize she's grabbed the prey, killing it instantly in her grip. She flies up to a snag broken off forty feet above the ground and tucks the mouse carefully into the jagged wood. This is a cache, not a nest. If she'd been delivering food to her young, the nest would be a natural platform high in a tree. She checks to be sure the mouse is well hidden. If she does have nestlings, she'll come back later for takeout.

The spotted owl research protocol demands that we spend an hour with the bird. She's had her limit of commercially raised albino mice, so now we sit to see what she does and if what she does will tell us whether she has a mate or nestlings.

The owl doesn't make a sound. She perches on a branch high above us. She is still. She watches us. She reaches her head forward—"the pre-pounce lean," Steve calls it—as if she has seen some prey on the ground. The song of a thrush flutters through the quiet, the auditory equivalent of seeing an orchid in the forest.

Beauty is what I came here for, a beauty enhanced, not diminished, by science. If I had only my senses to work with, how much thinner would be the experience. What a record we might have of the world's hidden beauty if field scientists and poets routinely spent time in one another's company.

A young tree, broken and caught between two others, creaks to the rhythm of the wind. How well the owl must know this sound. Does she anticipate the crash of its falling? What is the consciousness of a spotted owl? There she perches perceiving us, and here we sit perceiving her. We exchange the long, slow, interspecies stare—no fear, no threat, only the confusing mystery of the other. Steve knows her language well enough to speak a few words: the location call, a bark of aggression. Perhaps that means she thinks we are owls. We do not look like owls. But we do, briefly, behave like owls, catching and offering prey, being still, and turning our eyes to the forest.

"What are you?"

"What are you?"

That's the conversation we have with our eyes.

"What will you do next?"

"What will you do next?"

I keep falling into the owl's eyes. Then we stand up and hike down from that high place.

From *Field Notes*

THOMAS LOWE FLEISCHNER

Log Decomposition Reflections Plot

Some sites are just naturally richer with metaphor.

I step onto a downed log completely cloaked in vivid green moss. It gives way beneath my foot—the progress of mycelia and bacteria has proceeded enough that cavities of air lurk within the log.

I come here, to the palace of rot, to consider the recent passing of my dear friend Cathy. At her death, she was exactly one-tenth the age of these giant, still-alive trees.

Here, "death" and "life" are virtually indistinguishable: the downed logs more full of critters than when they were standing trees. Bright green mosses covering every surface on the ground and reaching upward into the trees. But the loss of Cathy's life—unlike the more nebulous "death" of these trees—seems to leave only a void. Not a nurturing seedbed, not a nurse log. Everyone around her feels emptiness, absence—none feels enriched.

return of the dead log people

JERRY MARTIEN

thank you for your participation in the blue river bone orchard's first
bicentennial morticultural conference: the role of the dead in carbon
budgeting. but don't think of it as over and done / we are everything
that is still to come

all the indignities you're afraid will happen to you
happen here

all the mortal invasions you keep from the house of the living /
from the porches of your ears / the eaves of your seeing / openings
you don't want eaten into / eggs hatching like little ideas in your brain /
microbes growing furry unspeakable words on your tongue /
the dark juices of your heart
gone to feed the living

upstart salal & Oregon grape
sapling of cedar & hemlock & fir
thriving in our cold wet breath

perceived by you as a chill in the air / in your green bones / those green bones
with which you thought you'd walk away from here
unchanged

but in your breath now
our breath

& in our breath
these words

which you will remember by a new / stiffness in your limbs
a whisper in your many-branched veins / & at last by

silence

& time

& your dust will rain on us / with the rain & we will take you in /
as easily as you breathed our air today

we eagerly await your input.

Denizens of Decay

TOM A. TITUS

A cashmere cowl of fog shrouding the night ridges thins and frays, retreating from morning sunshine. Light travels downward through the remaining vapor in angular rods that pry through green interlocking fingers of conifers whose arms are joined to furrowed trunks of dinosaur skin. The trees are so ancient and wise that to guess their age seems disrespectful. Yet even these Old Ones die. Their lives might end slowly from disease, leaving them stately and upright and home to wood-boring insects, in turn bringing woodpeckers who excavate cavities that become home to owls. Other giants are uprooted suddenly by Pacific storms, toppling with a muffled *WHUMP* to lie in state on the forest floor fully at the mercy of the incessant creeping moisture. The wetness brings a legion of microbes who feed on bark, sapwood, and heartwood, disassembling those who spent their lives as Owl Mother, Vole Lover, Shade Giver, and Protector of the Multitude of Greens. The crumbling interior of each log becomes a wet sponge, impervious to the penetrating shafts of sunlight, a dead host to many new lives.

Inside the log, an Oregon Slender Salamander lies tucked into a moist interstitial space where the bark has separated from the inner wood. She is pencil thin, no longer than my index finger, and has slipped easily into this tiny place. Her back is brick-red herringbone, the color of her heartwood home, her belly a flurry of white flakes in a dark sky, skin slightly swollen by a clutch of eggs beginning to accumulate yolk that is the first food for the next generation. In the wetness of spring, the salamander will spend her days inside this log, venturing forth only when darkness sends the

drying sun behind the ridges, keeping hungry garter snakes and birds at home.

Her beauty is painted onto a skin so thin that water travels effortlessly from salamander to forest and back again, so thin that oxygen made by green photosynthetic needles in the canopy moves easily into her body, so thin that carbon dioxide flows freely outward, her small gift of reciprocity to the trees. Her life depends upon this free exchange of gas and water from skin to forest and back again because she has no lungs. Her ancestors jettisoned these at least sixty million years ago, an act of evolutionary trust in which they committed themselves and their descendants to lives of abundant moisture and oxygen.

For Slender Salamander, there is vulnerability in this singular devotion to dampness. Her skin must remain moist and porous, and this openness is the very thing that causes her to lose water easily. She must adore humid darkness, revel in the press of her flanks against moist wood, delight in tiny bits of decay that cling to her head and legs as she negotiates the tight channels of her log. She must love this place. Because in summer when sunlight falls unimpeded from impossibly blue skies, when every footfall on the forest floor raises a resounding crunch, Slender Salamander senses certain death by desiccation. The womblike wetness within her log is a refuge from this seasonal version of climate disruption.

In spring, the long-dead Douglas-fir becomes a birthing ward. Deep within the interior of the log, sheltered from summer heat, Slender Salamander lays a cluster of six or so eggs, encircling them with the moist finger of her body, warding off predators and fungal infections. Her egg-bound young have long, feathery gills. But they will never feel a flowing spring or standing pond or any other water beyond the protective capsule of their own egg. Within this log womb they will hatch into tiny, fully metamorphosed versions of their parents.

This rebellion against the amphibian status quo of aquatic larvae has liberated Slender Salamander and her relatives in the lungless salamander family from dependence on freestanding water, allowing them to live and procreate on the vast slopes of moist forest in the western Cascades that are devoid of ponds and streams. Their revolt eliminated perilous spring breeding treks to water filled with aquatic predators, water that could disappear in the summer drought. Better the persistently damp safety of a large log.

Slender Salamander and her ancestors have hidden, eaten, and reproduced within these logs for perhaps twenty million years, far longer than humans have dwelt in these forests and an occupation incomprehensibly ancient compared with the mere two centuries since people began hauling trees from the ridges to become dry lumber for human houses rather than moist rot for salamander homes. Surely there is an accumulated intelligence in the longevity of slender salamanders, a way of living shaped by eons of evolution into a commitment to their place so complete that salamanders and logs have become singular and inseparable.

What would Slender Salamander teach us if we could squeeze in next to her, sit at her tiny, four-toed feet, and listen to her small breath that is not breathed? Maybe she would show us the difference between *timber*, a human construct measured in board feet, and *forests*, whose all-inclusive complexity defies measurement. We might discover the value of silence and how insight sometimes slips quietly through doors previously unrecognized. Perhaps we could learn to open our skin to the world.

GROUND WORK

Soundscape

Enter the forest. Be still. The quiet of the immense forest is startling, incongruous. Let the small sounds seep in. The continuous drone of a stream flowing over cobbles in riffles. A winter wren's song piercing the stream song. A Douglas squirrel's scolding chatter. A commercial jetliner streaming along the Cascade flight path. The whine of a truck on a nearby forest road. Geophony, biophony, anthrophony.

We tend to pick out sounds in the forest one by one. Birders are particularly attuned to and astute in gleaning information from sound. Over the past decade or so the field of soundscape ecology has taken shape in the global ecological science community. For those sounds audible to the human ear, researchers simply use their ears to sample; for those sounds pitched at frequencies human ears can't detect, such as the communicative, navigational, and locomotive sounds of bats, invertebrates, and marine animals, researchers use recording instrumentation. Avian ecologists have deployed sixteen listening devices and 188 listening stations across the Andrews Forest to detect birds by their songs and to assess their distribution across the mountain landscape as they migrate in response to seasonal shifts in microclimate and food sources. Computer scientists are aiding ecologists in this biodiversity assessment work by developing algorithms for distinguishing animal sounds from one another and from other sounds, and then identifying the species generating the sound by comparison with a reference library of songs. The challenges include the "cocktail party problem" of dealing with overlapping songs. Earth scientists sample variation in the geophonics of small and large streams in relation to distance from the stream and the role of streamside vegetation in muffling sound.

Still, we are very new to sensing and understanding the whole soundscape of the landscape. For example, do groups of bird species move around the landscape as communities in response to seasonal environmental shifts, or are species behaving individually? Does stream sound along the riparian network lacing through the landscape provide sonic cover from some species and deflect other species to quieter, upslope locations where stream noise doesn't complicate communications with friend and foe? These and other questions await new data and new tools.

Mind in the Forest

SCOTT RUSSELL SANDERS

I touch trees, as others might stroke the fenders of automobiles or finger silk fabrics or fondle cats. Trees do not purr, do not flatter, do not inspire a craving for ownership or power. They stand their ground, immune to merely human urges. Saplings yield under the weight of a hand and then spring back when the hand lifts away, but mature trees accept one's touch without so much as a shiver. While I am drawn to all ages and kinds, from maple sprouts barely tall enough to hold their leaves off the ground to towering sequoias with their crowns wreathed in fog, I am especially drawn to the ancient, battered ones, the survivors.

Recently I spent a week in the company of ancient trees. The season was October and the site was the drainage basin of Lookout Creek, on the western slope of the Cascade Mountains in Oregon. Back in my home ground of southern Indiana, the trees are hardwoods—maples and beeches and oaks, hickories and sycamores—and few are allowed to grow for as long as a century without being felled by ax or saw. Here, the ruling trees are Douglas-firs, western hemlocks, western redcedars, and Pacific yews, the oldest of them ranging in age from five hundred to eight hundred years, veterans of countless fires, windstorms, landslides, insect infestations, and floods.

* * *

On the first morning of my stay, I follow a trail through moist bottomland toward Lookout Creek, where I plan to spend half an hour or so in meditation. The morning fog is thick, so the treetops merge with gray sky. Condensation drips from every needle and leaf. My breath steams. Lime-green lichens, some as long as a horse's tail, dangle from branches. Set off against the somber greens and browns of the conifers, the yellow and red leaves of vine maples, big-leaf maples, and dogwoods appear luminous in spite of the damp. Shelf fungi jut from the sides of old stumps like tiny balconies, and hemlock sprigs glisten atop nurse logs. The undergrowth is as dense as a winter pelt.

Along the way, I reach out to brush my fingers over dozens of big trees, but I keep moving, intent on my destination. Then I come upon a Douglas-fir whose massive trunk, perhaps four feet in diameter at chest height, is surrounded by scaffolding, which provides a stage for rope-climbing by scientists and visiting schoolchildren. Something about this tree—its patience, its generosity, its dignity—stops me. I place my palms and forehead against the furrowed, moss-covered bark and rest there for a spell. Gradually the agitation of travel seeps out of me and calm seeps in. Only after I stand back and open my eyes, and notice how the fog has begun to burn off, do I realize that my contact with this great tree must have lasted fifteen or twenty minutes.

I continue on to a gravel bar on Lookout Creek, a jumble of boulders, cobbles, pebbles, and grit scoured loose from the volcanic plateau that forms the base of the Cascade Mountains. Because these mountains are young, the slopes are steep and the water moves fast. Even the largest boulders have been tumbled and rounded. Choosing one close to a riffle, I sit cross-legged and half close my eyes, and I am enveloped in water sounds, a ruckus from upstream and a burbling from downstream. Now and again I hear the thump of a rock shifting in the flow, a reminder that the whole mountain range is sliding downhill, chunk by chunk, grain by grain.

Although I have tried meditating for shorter or longer stretches since my college days, forty years ago, I have never been systematic about the practice, nor have I ever been good at quieting what Buddhists call the "monkey mind." Here beside Lookout Creek, however, far from my desk and duties, with no task ahead of me but that of opening myself to this place, I settle quickly. I begin by following my breath, the oldest rhythm of flesh, but soon I am following the murmur of the creek, and I am gazing at the bright leaves of maples and dogwoods that glow along the thread of the stream like jewels on a necklace, and I am watching light gleam on water shapes formed by current slithering over rocks, and for a spell I disappear, there is only this rapt awareness.

* * *

Each morning at first light I repeat the journey to Lookout Creek, and each time I stop along the way to embrace the same giant Douglas-fir, which smells faintly of moist earth. I wear no watch. I do not hurry. I stay with the tree until it lets me go.

When at length I lean away, I touch my forehead and feel the rough imprint of the bark. I stare up the trunk and spy dawn sky fretted by branches. Perspective makes the tops of the surrounding, smaller trees appear to lean toward this giant one, as if conferring. The cinnamon-colored bark is like a rugged landscape in miniature, with flat ridges separated by deep fissures. Here and there among the fissures, spider webs span the gaps. The plates are furred with moss. A skirt of sloughed bark and fallen needles encircles the base of the trunk. Even in the absence of wind, dry needles the color of old pennies rain steadily down, ticking against my jacket.

I don't imagine that my visits mean anything to the Douglas-fir. I realize it's nonsensical to speak of a tree as patient or generous or dignified merely because it stands there while researchers and children clamber up ropes into its highest limbs. But how can I know a tree's inwardness? Certainly there is intelligence here, and in the forest as a whole, if by that word we mean the capacity for exchanging information and responding appropriately to circumstances. How does a tree's intelligence compare with ours? What can we learn from it? And why, out of the many giants thriving here, does this one repeatedly draw me to an embrace?

The only intelligence I can examine directly is my own and, indirectly, that of my species. We are a contradictory lot. Our indifference to other species, and even to our own long-term well-being, is demonstrated everywhere one looks, from the depleted oceans to the heating atmosphere, from poisoned wetlands to eroding farmlands and forests killed by acid rain. Who can bear in mind this worldwide devastation and the swelling catalogue of extinctions without grieving? And yet it's equally clear that we are capable of feeling sympathy, curiosity, and even love toward other species and toward the Earth. Where does this impulse come from, this sense of affiliation with rivers and ravens, mountains and mosses? How might it be nurtured? What role might it play in moving us to behave more caringly on this beleaguered planet?

These are the questions I find myself brooding about as I sit in meditation beside Lookout Creek. One is not supposed to brood while meditating, of course, so again and again I let go of thoughts and return my awareness to the water sounds, the radiant autumn leaves, the wind on my cheek, the stony cold chilling my sitting bones. And each morning, for shorter or longer spells, the fretful *I* quiets down, turns transparent, vanishes.

Eventually I stir, roused by the haggle of ravens or the chatter of squirrels or the scurry of deer—other minds in the forest—and I make my way back along the trail to the zone of electricity and words. As I walk, it occurs to me that meditation is an effort to become, for a spell, more like a tree, open to whatever arises, without judging, without remembering the past or anticipating the future, fully present in the moment. The taste of that stillness refreshes me. And yet I do not aspire to dwell in such a condition always. For all its grandeur and beauty, for all its half-millennium longevity, the Douglas-fir cannot ponder me, cannot reflect or remember or imagine—can only *be*. Insofar as meditation returns us to that state of pure, unreflective being, it is a respite from the burden of ceaseless thought. When we surface from meditation, however, we are not turning from reality to illusion, as some spiritual traditions would have us believe; we are reclaiming the full powers of mind, renewed by our immersion in the realm of mountains and rivers, wind and breath.

* * *

At midday, sunlight floods the gravel bar on Lookout Creek, illuminating strands of spider filament that curve from one boulder to another over an expanse of rushing water. At first I can't fathom how spiders managed this engineering feat. The wind might have blown them one direction but not back again, and yet at least a dozen gossamer threads zigzag between the massive stones. Then I guess that the spiders, after attaching the initial strand, must climb back and forth, adding filaments. The stones they stitch together are as knobby and creased as the haunches of elephants. Even in still air, butter-yellow maple leaves come sashaying down. A pewter sheen glints from the bark of young Douglas-firs tilting out over the stream.

Unconsciously, I resort to human terms for describing what I see, thus betraying another quirk of our species. We envision bears and hunters and wandering sisters in the stars. We spy dragons in the shapes of clouds, hear mournfulness in the calls of owls. Reason tells us that such analogies are false. For all its delicious sounds, the creek does not speak, but merely slides downhill, taking the path of least resistance, rubbing against whatever it meets along the way. Boulders have nothing to do with elephants, lichens are not horsetails, moss is not fur, spiders are not engineers, ravens do not haggle, and trees do not confer. Scientists are schooled to avoid such anthropomorphism. Writers are warned against committing the "pathetic fallacy," which is the error of projecting human emotions or meanings onto nature. The caution is worth heeding. Yet if we entirely forgo such analogies, if we withhold our metaphors and stories, we estrange ourselves from the universe. We become mere onlookers, the sole meaning-bearing witnesses of a meaningless show.

But who could sit here, on this gravel bar beside Lookout Creek, and imagine that we are the sole source of meaning? Against a halcyon blue sky, the spires of trees stand out with startling clarity, their fringe of lichens appearing incandescent. Moths and gnats flutter above the stream, chased by dragonflies. The creek is lined by drift logs in various states of decay, from bone-gray hulks to rotting red lumps. Wet boulders gleam as if lit from within. Cobbles jammed against one another look like the heads of a crowd easing downstream. The muscular current, twisting over rocks, catches and tosses the light. The banks on either side blaze with the salmon-pink leaves of dogwoods, those western relatives of the beloved understory tree of my Indiana forests. Everything in sight is exquisite—the stones of all sizes laid against one another just so, the perforated leaves of red alders, the fallen needles gathered in pockets along the shore, the bending grasses, the soaring trees.

Only cosmic arrogance tempts us to claim that all this reaching for sunlight, nutrients, and water means nothing except what we say it means. But if it bears a grander significance, what might that be, and what gives rise to such meaning? What power draws the elements together and binds them into a spider or a person, a fern or a forest? If we answer, "Life," we give only a name, not an explanation.

* * *

Those who fancy that humans are superior to the rest of nature often use "tree hugger" as a term of ridicule, as if to feel the allure of trees were a perverted form of sensuality or a throwback to our simian ancestry. Of course, many who decry tree-hugging don't believe we *have* a simian ancestry, and so perhaps what they fear is a reversion to paganism. And they may have a point. The religions that started in the Middle East—Judaism, Christianity, Islam—are all desert faiths, created by people who lived in the open. Theirs is a sky god, who would be eclipsed by a forest canopy. In every civilization influenced by these faiths, trees have been cut down not merely to secure wood for cooking and building or to clear ground for agriculture or to open vistas around settlements where predators might lurk, but to reveal the heavens.

Worship of a sky god has been costly to our planet. Religions that oppose the heavenly to the earthly, elevating the former and scorning the latter, are in effect denying that we emerge from and wholly depend on nature. If you think of the touchable, eatable, climbable, sexy, singing, material world as fallen, corrupt, and sinful, then you are likely to abuse it. You are likely to say that we might as well cut down the last old-growth forests, drain the last swamps, catch the last tuna and cod, burn the last drops of oil, since the end time is coming, when the elect few will be raptured away to the immortal realm, and everything earthly will be utterly erased.

But our language preserves a countervailing wisdom. In Latin, *materia* means "stuff," anything substantial, and in particular it means "wood." *Materia* in turn derives from *mater*, which means "mother." In the collective imagination that gave rise to these meanings, trees were understood to epitomize matter, and matter was understood to be life-giving. Perhaps we could tap into this wisdom by recovering another word that derives from *mater: matrix*, which means "womb." Instead of speaking about *nature* or *the environment,* terms that imply some realm apart from us, perhaps we should speak of Earth as our matrix, our mother, the source and sustainer of life.

It is easy to feel nurtured among these ancient trees. I breathe the forest. I drink its waters. I take in the forest through all my senses. In order to survive here for any length of time, I would need to wear the forest, its fur and skin and fiber; I would need to draw my food from what lives here alongside me; I would need to burn its fallen branches for cooking and for keeping warm; I would need to frame my shelter with its wood and clay and stone. Above all, I would need to learn to *think* like the forest, learn its patterns, obey its requirements, align myself with its flow.

There are no boundaries between the forest and the cosmos, or between myself and the forest, and so the intelligence on display here is continuous with the intelligence manifest throughout the universe and with the mind I use to apprehend and speak of it.

* * *

One morning beside Lookout Creek, enveloped as usual in watery music, I sit leaning against a young red alder that has sprouted in the gravel bar, its leaves nibbled into lace by insects. Everything here either starts as food or winds up as food. None of the alders growing on this ever-shifting bank is thicker than a baseball bat. The next big flood will scour them away. Beside me, the sinewy roots of an upturned stump seem to mimic the muscular current in the stream. The bar is littered with gray and ruddy stones pockmarked by holes that betray the volcanic origins of this rubble.

Where better than such a place to recognize that the essence of nature is *flow*—of lava, electrons, water, wind, breath. *Materia*, matter, the seemingly solid stuff we encounter—trees, stones, bears, bones—is actually fluid, constantly changing, like water shapes in the current. The Psalmist tells us, "The mountains skipped like rams, and the little hills like lambs," and Dōgen, a thirteenth-century Zen teacher, proclaims that mountains are always walking. Both speak truly. Mountains do move, arising and eroding away over geological time, just as organisms grow and decay, species evolve, tectonic plates shift, stars congeal and burn and expire, entire galaxies shine for a spell and then vanish. Nothing in nature is fixed.

Conservationists have often been accused of wishing to freeze the land in some favored condition—for example, the American continent as it was before European colonization. Back when maps described old growth as large saw-timber, scientists spoke of forests reaching climax, as if at some point the flow would cease. But we now realize that no such stasis is possible, even if it were desirable. If flux is the nature of nature, however, we still must make distinctions among the *kinds* of change. We cannot speak against the damage caused by human behavior unless we distinguish between *natural* change—for example, the long history of extinctions—and *anthropogenic* change—for example, the recent acceleration in extinctions due to habitat destruction, pollution, climate change, and other disturbances caused by humans. The capacity to make such a distinction, and to act on it, may be as unique to our species as the capacity to use symbolic language.

Thoughts flow, along with everything else, even in the depths of meditation. And yet the human mind seems compelled to imagine fixity—heaven, nirvana, Plato's ideal realm, eternal God—and the human heart yearns for permanence. Why else do we treasure diamonds and gold? Why else do Creationists cling to the notion that all species were made in exactly their present form? Why else do we search for scientific "laws" underlying the constant flux of the universe?

Our yearning for the fixed, like our craving for dominion over nature, may be another expression of our fear of aging and death. This occurs to me as I sit, transfixed, beside the narrowest, noisiest passage in the riffles on Lookout Creek. A dozen snags tilt above my head, their bare limbs like the sparse whiskers on an old man's chin. Upstream, a gigantic Douglas-fir has fallen across the creek, its trunk still as straight as when it was alive. Downstream, another giant has fallen, this one snapped in the middle. I can't help imagining one of the looming snags suddenly toppling onto me and snapping my thread of thought, scattering this congregation of elements and notions bearing my name.

* * *

Higher up the valley of Lookout Creek, in a grove of five-hundred-year-old Douglas-firs and western hemlocks, a hundred or so logs have been placed side by side on the ground, labeled with aluminum tags, and fitted with instruments to measure their rate and manner of decay. Designed to continue for two centuries, this research aims to document, among other things, the role of dead wood in forest ecology and in the sequestering of carbon.

On a visit to the site, I stroke the moss-covered logs, touch the rubbery fungi that sprout from every surface, peer into the boxy traps that catch flying insects and fallen debris, and lean close to the tubes that capture the logs' exhalations. The only breathing I detect is my own. I'm intrigued that scientists are studying decomposition, for as an artist I usually think about *composition*—the making of something shapely and whole out of elements. A musician composes with notes, a painter with colors, a writer with letters and words, much as life orchestrates carbon, oxygen, nitrogen, and other ingredients into organisms. These organisms—trees, fungi, ravens, humans—persist for a while, change over time, and eventually dissolve into their constituents, which will be gathered up again into living things.

Art and life both draw energy from sunlight, directly or indirectly, to counter entropy by increasing order. Right now, for example, I'm running on the sunshine bound up in pancakes and maple syrup. Organisms interact biophysically with everything in their ecosystem and, ultimately, with the whole universe. By contrast, the symbolic structures that humans create—songs, stories, poems, paintings, photographs, films, diagrams, mathematical formulas, computer codes—convey influence only insofar as they are read, heard, or otherwise perceived by humans. What happens when we turn our interpretive powers on living organisms? Does raven, Douglas-fir, spider, or lichen mean anything different, or anything more, when it is taken up into human consciousness?

What we think or imagine about other species clearly influences our behavior toward them—as notions about the wickedness of wolves led to their extermination throughout much of their historic range, and as new understanding about the role of predators has led to the reintroduction of wolves in Yellowstone and elsewhere. But aside from this practical impact, does our peculiar sort of mind bear any greater significance in the scheme of things? Is it merely an accidental result of mechanical processes, an adaptive feature that has powered our—perhaps fleeting—evolutionary success? Would the universe lose anything vital if our species suddenly vanished?

We can't know the answer to those questions, despite the arguments of prophets and philosophers. We can only form hunches, and, right or wrong, these will influence the spirit of our work and the tenor of our lives. For what it's worth, my hunch is that what we call mind is not a mere side effect of material evolution, but is fundamental to reality. It is not separate from what we call matter, but is a revelation of the inwardness of things. I suspect that our symbol-wielding intelligence is a manifestation of the creative, shaping energy that drives the cosmos, from the dance of electrons to the growth of trees. If this is so, then our highest calling may be to composition—paying attention to some portion of the world, reflecting on what we have perceived, and fashioning a response in words or numbers or paint or song or some other expressive medium. Our paintings on cave walls, our photos of quasars, our graphs and sonnets and stories may be the gifts we return for the privilege of sojourning here on this marvelous globe.

* * *

If intelligence means the ability to take in and respond to information, then all organisms possess it, whether animal or plant, for they exchange signals and materials with their surroundings constantly. If intelligence means the capacity for solving puzzles or using language, then surely the ravens that clamor above me and the wolves that roam the far side of the mountains possess it. But if we are concerned with the power not merely to reason or use language, but to discern and define meanings, to evaluate actions in light of ethical principles, to pass on knowledge across generations through symbolic forms—then we are speaking about a kind of intelligence that appears to be the exclusive power of humans, at least on this planet.

Some contemplative traditions maintain that this meaning-making capacity is a curse, that it divorces us from reality, enclosing us in a bubble of abstractions. It's easy to sympathize with this view, when one considers our history of feuds and frauds. Cleverness alone does not make us wise. Yet here among these great trees and boisterous mountain streams, I sense that our peculiar sort of mind might also be a blessing, not only to us but also to the forest, to other creatures, to life on Earth, and even to the universe.

I recognize the danger of hubris. It's flattering to suppose, as many religions do, that humans occupy a unique place in the order of things. The appeal of an idea is not evidence for its falsity, however, but merely a reason for caution. Cautiously, therefore: Suppose that the universe is not a machine, as nineteenth-century science claimed, but rather a field of energy, as twentieth-century science imagined. Suppose that mind is not some private power that each of us contains but rather a field of awareness that contains us—and likewise encompasses birds, bees, ferns, trees, salamanders, spiders, dragonflies, and all living things, permeates mountains and rivers and galaxies, each kind, offering its own degree and variety of awareness, even stars, even stones.

What if our role in this all-embracing mind is to gaze back at the grand matrix that birthed us and to translate our responses into symbols? What if art, science, literature, and our many other modes of expression feed back into the encompassing mind, adding richness and subtlety? If that is our distinctive role, no wonder we feel this urge to write, to paint, to measure and count, to set strings vibrating, to tell stories, to dance and sing.

Coda

VICKI GRAHAM

From the larger poem *Debris*

The poet asks: What is the smallest bit of matter?
and the scientist replies: A string.

String.   Knot.   Web.   Watershed.

Only in the mind can there be
one word, one note, one brushstroke;
only in the mind can there be one orchid,
one owl, one vole, one fungus, one fir.

Only in the mind can there be one forest.

compose   decompose
pause
recompose

Afterword

Advice to a Future Reader

KATHLEEN DEAN MOORE

To a Future Reader:

The audacity of the Long-Term Ecological Reflections project makes me grin. To set out to compile, over a span of two hundred years, an unbroken chronicle of creative responses to the forest? The longevity of the whole thing intrigues me. I can imagine a future reader, two hundred years from today, reading backward through the record and arriving, finally, at now. I don't know if you will be sitting in a forest or floating in a plastic bubble. I don't know if the stories will seem otherworldly to you, bittersweet fables of a vanished land viewed through a lens that has long ago cracked and clouded. Or maybe not: Maybe you will be glad to recognize in these stories the same green and light-graced forests that you know and love, and the same wonder at their beauty. Or maybe you will be surprised by how paltry and unprotected our forests are, compared to your own green-glorious wildlands, and how impoverished our understanding is. I wish I knew.

I wish I could invite you back in time to talk. You would find me in a clearing in the forest, on the porch of the research station. I will be wearing a blue fleece jacket made from recycled plastic milk jugs and a baseball cap that says, "Life is good." I would pour you black tea from India and pull open a can of sardines (do you know sardines?). We would sit on the porch, by the hummingbird feeder (do you know hummingbirds?) and look past the alders to the great trees. I would give you this book, this record of the first decade of twenty decades, and then I would bring my chair close to yours and explain my ideas about how to read it.

1. THE WORDS

Pay attention to the words, I will say. They will tell you what the writers see and don't see, and what they think is important and what is not. The words are a record of *what comes to mind*—literally that, which is a *worldview*; literally that, a vision of the world. So what are the words of this decade? I asked my computer to count them, which may seem quaint to you, or miraculous, depending on how things go with the world. These words are used a hundred times or more:

forest trees growth
time years water
long moss science
owl life

These appear a dozen times:

language fallen
future deep
white nitrogen blue death
fungi why

And some words make cameo appearances:

woodrat worship
wasp blessing
Tupperware Thoreau
unfurling unfathomable

A word count is no substitute for understanding, I'm aware; I know that a good writer will not often name the idea she wants to express, letting licorice fern or tree-ruffle liverwort speak of life returning, letting drifts of light speak for wonder or the possibility of photosynthesis on the forest floor.

But I have read these essays and poems carefully, and that reading confirms what the word count suggests—that in this decade, the forest has come into its own. Its meaning is not deduced from human ends. It does not stand for something greater or reach for some human-imposed ideal. It is what it is. And what it is, is a system—the "tangled banks," the red-spotted specificity of mutually sustaining relationships. Everything is included in the system, even the dead stuff, even *Homo sapiens*. This is significant—the uncomprehending seamlessness of all lives, including our own. The human being is of the whole, not metaphorically, but chemically and culturally. *Forest* is now the word for the process of weaving that makes an endless tapestry, tearing and mending, weaving old and new, creating a complex forest, where each watery warp and fledgling woof, each part, is astonishing and alive and worthy, worthy, the stump and the stone, the rain, the wasp.

2. THE VOICES

Pay attention to who is speaking and who is listening. Pay attention to what is heard and what is not. The poets and essayists are listening intently in this forest. Picture them: they stand quietly in a mossy opening surrounded by three-hundred-year-old hemlocks or find a quiet listening seat—a damp log over a river, a coffin-shaped space beneath a fallen tree, a grove carpeted in sorrel, a stump in an old scar.

Ask, what do they hear in that time and in that place? What are the stories that ring out, repeated like thrush calls in the morning? In this decade, I believe they hear these:

> *Time itself, the expanse and slide of it.*
> Not the hoot of an owl on a single snowy night, but the
> hooting of all nights, extended into a jagged line drawn
> on a chart—these many tonight, that many last night. The

movement through time of a mountain slope, of the creek, of the leaders on a thousand Douglas-firs, all these lines on all these charts, tangled through time, like music that had been there all along, waiting for someone who can transpose it into a single song.

Not only life, but also death and the dead and the undead, an entire zombie world.
Downed trees on the forest floor, the muffled scream of a red vole in the gullet of an owl, duff and litter and fallen needles, dead wood in streams, hectares of rock and soil come back to life, sliding slowly downslope. The life in this death, the death in this life—nestlings, saplings, fingerling trout.

The complex and relational.
The meaning of an entity, in fact its very being, is found here in complex relation to other things—not an isolated pile of dimensional lumber, not a princely Douglas-fir, but blistered bark and fire and lichen and politics and the scrabbling claws of fledgling owls through time and the rain of needles. In this decade, the voices of interdependence sing out—in increasing numbers, the voices of women. Here are poets, scientists, elders, measurers of air, gatherers of berries, taggers of fish, tellers of stories, tenders of the next generations—people who define themselves by their relationships to other beings and notice when a living tree has reached out its roots to nourish a still-seeping stump.

Invisible things.
Things that sail through the night, things as thin as spiderwebs, spiderwebs themselves, buried things, submerged things, rivers under rivers, rivers over rivers, concealed mycorrhizomes. Air, moving and still.

Radioactivity in the rain. The hidden past, but little
mention of the future.

Ask, who speaks with the loudest voices? The machines are muffled in this book, although they are an angry mutter in the background. The rain speaks. Returning life sings out. But if these essays represent the decade, it's scientists who are most clearly heard. Poets follow scientists with open notebooks and eager pens, hungry for the words, hungry for the close encounters. Even when the scientists are gone back to their university offices, their presence is noisy in the forest, in the pink flags and white buckets, their experimental plots. With their exquisitely tuned ears, poets sit beside the pink flags and hear them speak.

Ask, what is silent in this creative account of the forest? There is no Cassandra; not yet. The voices of future generations are only a murmur, uneasy in the belly of the pregnant world. When it comes to the cries of the crowded cities and the roar of coal and gas on fire, the forest is still silent, perhaps one of the last remaining silent places on Earth, not speaking of ruin or salvation. Not yet.

3. THE SHAPE

Pay attention to the shape of the story. That's what makes a story, right?—it has a form, a beginning, a middle, and an end. What is the shape of the chronicles in this book?

If I were to draw an image of the decades *before* the time of this book, I might dip my Chinese ink brush into black ink and drag a circle around the page, counterclockwise from the top. But the brush would stop just before the circle was closed and jag off toward the top right, ending in an arrow that points forever into limitless progress. This was the era of ferocious logging, fueled by a belief in endless advancement, dollars that grow on trees, jobs and profits that reproduce *themselves*, by God. Start with the dark, fecund cycles of birth and growth, and at the apex of growth, harvest, harvest, and saw these big logs into money. Interesting, that this shape is also the symbol for the male; those were deep-voiced, diesel-driven decades, no doubt.

The writers in the decade of this book draw a different picture. Dip the brush in black ink. Start at the top and pull the brush around in a circle. Close the circle and keep on following the curve, around and up again and down. As the ink fades, dip it again into sun and rain and rotting roots and keep on circling until the cycle is dripping and impenetrable—a shiny wet eternity of rise and return.

But I think—although I cannot be sure—that I see a small change in the picture the writers will paint toward the end of this decade. These writers dip the brush in the same fecund ink and draw a thick circle. But the next circle is slightly displaced, and the next displaced again—not much, only the width of a hemlock's needle. Nevertheless, the circle is a spiral now, the forest never returning to the same condition, the image winding its energy like a spring.

I will guess (and you in the future can tell me if I'm right or wrong) that the writers of the next decades will draw the tightly wound spiral, but at some time—soon, maybe; later, most likely—the brush will veer from the spiral, as if a spring had jumped loose, and the brush will leap off in wild directions, dragging the line of ink behind it. And then the dense circle of repeating, reassuring rhythms will be all anyone wants to write about, an object of loss and wondering love.

* * *

I'm going to take my book and my cold tea and go inside. I've stayed on the porch too long, and now my knees are wet. A gale has poured in from the ocean, tumbling the crows. Hemlock boughs rise up and down, as if they are patting the alders on the head. The fledgling thrushes have stopped practicing their bent whistles; fluffed and silent, they cling miserably to the leeward branches of a big-leaf maple. Rain jets through in pulses, as if shot from a kinked hose. I think you will know about gales. I'm not sure you will know crows or thrushes. You will know rain. I think you will know how I feel: cold and wishing for dry clothes, worried about the treetops bucking and slapping, cheered by the flap-dance of the alders, wondering what it all means, and hoping, hoping that you are here two hundred years from now, and that you are well.

For Further Reading

We hope this book will serve as an invitation to readers to further explore the creative writing, art, and science of the Long-Term Ecological Reflections program and the Andrews Forest. The Spring Creek Project website (http://springcreek.oregonstate.edu/) has more information about the Reflections program and additional references, plus the complete collection of contributions to the *Forest Log*, from which this book has been drawn.

BRIDGING SCIENCE AND HUMANITIES

Books

Goodrich, Charles, Kathleen Dean Moore, and Frederick J. Swanson, eds. *In the Blast Zone: Catastrophe and Renewal on Mount St. Helens*. Corvallis: Oregon State University Press, 2008.

Gould, Stephen J. *The Hedgehog, the Fox, and the Magister's Pox: Mending the Gap between Science and the Humanities.* New York: Harmony Books, 2003.

Haskell, David George. *The Forest Unseen: A Year's Watch in Nature.* New York: Viking Penguin, 2012.

Nabhan, Gary Paul. *Cross-Pollinations: The Marriage of Science and Poetry.* Minneapolis, MN: Milkweed Editions, 2004.

Preston, Jane, ed. *The Language of Conservation: Poetry in Library and Zoo Collaborations.* New York: Poets House, 2013.

Wilson, Edward O. *Consilience.* New York: Knopf, 1998.

Essays and Articles

Deming, Alison Hawthorne. "Science and Poetry: A View from the Divide." *Creative Nonfiction* 11 (1998): 11–29.

Elder, John, and Glenn Adelson. "Robert Frost's Ecosystem of Meanings in 'Spring Pools.'" *ISLE* 13, no. 2 (Summer 2006): 1–17.

Swanson, Frederick J., Charles Goodrich, and Kathleen Dean Moore. "Bridging Boundaries: Scientists, Creative Writers, and the Long View of the Forest." *Frontiers in Ecology and the Environment* 6, no. 9 (2008): 449–504.

Website

Ecological Reflections network: www.ecologicalreflections.com/.

ANDREWS FOREST ECOLOGY AND NATURAL/CULTURAL HISTORY

Books

Durbin, Kathie. *Tree Huggers: Victory, Defeat, and Renewal in the Northwest Ancient Forest Campaign*. Seattle, WA: Mountaineers, 1996.

Franklin, J. F., and C. T. Dyrness. *Natural Vegetation of Oregon and Washington*. Corvallis: Oregon State University Press, 1988.

Geier, Max G. *Necessary Work: Discovering Old Forests, New Outlooks, and Community on the H. J. Andrews Experimental Forest, 1948–2000*. General Technical Report PNW-GTR-687. Portland, OR: U.S. Department of Agriculture, Forest Service, Pacific Northwest Research Station, 2007.

Hays, Samuel P. *Wars in the Woods: The Rise of Ecological Forestry in America*. Pittsburgh, PA: University of Pittsburgh Press, 2007.

Luoma, Jon R. *The Hidden Forest: The Biography of an Ecosystem*. Corvallis: Oregon State University Press, 2006.

Spies, Thomas A., and Sally L. Duncan, eds. *Old Growth in a New World: A Pacific Northwest Icon Reexamined*. Washington, DC: Island Press, 2009.

RESEARCH TOPICS REFERENCED IN THE BOOK

"Can There Be Orderly Harvest of Old Growth?" *Timberman* 58, no. 10 (1957): 48–52. http://andrewsforest.oregonstate.edu/pubs/biblio/abstract.cfm?Catalog_id=1104&topnav=175.

Cissel, John H., Frederick J. Swanson, and Peter J. Weisberg. "Landscape Management Using Historical Fire Regimes: Blue River, Oregon." *Ecological Applications* 9, no. 4 (1999): 1217–1231.

Forsman, Eric D., and others. *Population Demography of Northern Spotted*

Owls. Studies in Avian Biology, no. 40. Berkeley, CA: University of California Press, 2011. A publication of the Cooper Ornithological Society.

Franklin, Jerry F., Kermit Cromack Jr., William Denison, Arthur McKee, Chris Maser, James Sedell, Fred Swanson, and Glen Juday. *Ecological Characteristics of Old-Growth Douglas-Fir Forests.* General Technical Report PNW-118. Portland, OR: U.S. Department of Agriculture, Forest Service, Pacific Northwest Forest and Range Experiment Station, 1981.

Halpern, C. B., and J. A. Lutz. "Canopy Closure Exerts Weak Controls on Understory Dynamics: A 30-Year Study of Overstory-Understory Interactions." *Ecological Monographs* 83 (2012): 221–237.

Harmon, M. E., J. F. Franklin, F. J. Swanson, P. Sollins, S. V. Gregory, J. D. Lattin, N. H. Anderson, S. P. Cline, N. G. Aumen, J. R. Sedell, G. W. Lienkaemper, K. Cromack Jr., and K. W. Cummins. "Ecology of Coarse Woody Debris in Temperate Ecosystems." In *Advances in Ecological Research,* ed. A. MacFadyen and E. D. Ford, 15:133–302. Orlando, FL: Academic Press, 1986.

Jones, J. A., G. L. Achterman, L. A. Augustine, I. F. Creed, P. F. Folliott, L. MacDonald, and B. C. Wemple. "Hydrologic Effects of a Changing Forested Landscape—Challenges for the Hydrological Sciences." *Hydrological Processes* 23 (2009): 2699–2704.

Morrison, Peter H., and Frederick J. Swanson. *Fire History and Pattern in a Cascade Range Landscape.* General Technical Report PNW-GTR-254. Portland, OR: U.S. Department of Agriculture, Forest Service, Pacific Northwest Research Station, 1990.

Swanson, Frederick J., Sherri L. Johnson, Stanley V. Gregory, and Steven A. Acker. "Flood Disturbance in a Forested Mountain Landscape." *BioScience* 48, no. 9 (1998): 681–689.

Swanson, Mark E., Jerry F. Franklin, R. L. Beschta, Charles M. Crisafulli, Dominick A. DellaSala, Richard L. Hutto, David B. Lindenmayer, and Frederick J. Swanson. "The Forgotten Stage of Forest Succession: Early-Successional Ecosystems on Forest Sites." *Frontiers in Ecology and the Environment* 9, no. 2 (2010): 917–925.

Waring, R. H., and J. F. Franklin. "Evergreen Coniferous Forests of the Pacific Northwest." *Science* 204 (1979): 1380–1386.

Winkel, G. "When the Pendulum Doesn't Find Its Center: Environmental Narratives, Strategies, and Forest Policy Change in the US Pacific Northwest." *Global Environmental Change* 27 (2014): 84–95.

WEBSITE

Andrews Forest science publications: http://andrewsforest.oregonstate.edu/lter/pubs.cfm?frameURL=175.

About the Editors

Nathaniel Brodie received his master of fine arts degree in nonfiction creative writing from the University of Arizona, where he served as the Beverly Rodgers Fellow and as a 1985 Graduate Fellow in the Arts and Humanities. His essays have appeared or are forthcoming in a number of magazines, literary journals, and anthologies—including *Creative Nonfiction*, *High Desert Journal*, *High Country News*, and *Terrain.org*. He was the recipient of the PEN Northwest 2014 Margery Davis Boyden Wilderness Writing Residency and was a finalist for both the 2013 Ellen Meloy Desert Fund and the 2015 Waterston Desert Writing Prize.

Charles Goodrich, director of the Spring Creek Project for Ideas, Nature, and the Written Word, is the author of three volumes of poems—*Insects of South Corvallis*, *Going to Seed: Dispatches from the Garden*, and *A Scripture of Crows*—and a collection of essays about nature, parenting, and building his own house, *The Practice of Home*. He coedited *In The Blast Zone: Catastrophe and Renewal on Mount St. Helens*. His essays and poetry have appeared in many magazines, including *Orion*, *Willow Springs*, *Zyzzyva*, *The Sun*, and *Best Essays Northwest;* and a number of his poems have been read by Garrison Keillor on the National Public Radio program *The Writer's Almanac*. Goodrich has a master of fine arts degree in creative writing from Oregon State University.

Frederick J. Swanson is an emeritus scientist with the Pacific Northwest Research Station of the U.S. Forest Service, a professor (courtesy) in the Department of Forest Ecosystems and Society at Oregon State University, and a senior fellow with the Spring Creek Project. His highly interdisciplinary science experience has been based at the Andrews Forest since 1972, and he has collaborated intensively with the humanities community over the past dozen years. Recent publications include *Bioregional Assessments: Science at* *the Crossroads of Management and Policy*; *Ecological Responses to the Eruption of Mount St. Helens;* and *In the Blast Zone*. He holds a PhD in geology from the University of Oregon.

About the Contributors

Sandra Alcosser's poems have appeared in the *New Yorker, New York Times, Paris Review, Ploughshares, Poetry,* and the *Pushcart Prize Anthology. A Fish to Feed All Hunger* and *Except by Nature* received the highest awards from the National Poetry Series, Academy of American Poets, and Associated Writing Programs. She serves on the graduate faculty of Pacific University and San Diego State University.

Kristin Berger is the author of *For the Willing* (Finishing Line Press, 2008). She is the recipient of residencies at the H. J. Andrews Experimental Forest, Starkey Experimental Forest and Range, and Playa at Summer Lake. Her current work appears in, or is forthcoming from, *Arc Poetry Magazine, Mi-POesias, Poecology, Terrain.org, Wayfarer Journal,* and *Written River.* Visit Kristin's website at Kristinberger.wordpress.com.

James Bertolino taught literature and creative writing for thirty-six years and retired from a position as writer in residence at Willamette University in 2006. His twelfth volume of poetry, *Ravenous Bliss: New and Selected Love Poems,* was published by MoonPath Press in 2014. His books have issued from presses at Princeton, Cornell, Brown, and Carnegie-Mellon Universities.

Joseph Bruchac's work often reflects his American Indian (Abenaki) ancestry and his lifelong commitment to the preservation of the natural world. He is author of over 120 books for young readers and adults, including *Our Stories Remember: American Indian History, Culture, and Values through Storytelling,* and the Keepers of the Earth series.

John R. Campbell's essays and poems have appeared in many literary journals, including *Poetry, Georgia Review,* and *North American Review.* His book of environmental meditations, *Absence and Light,* is available from the University of Nevada Press. Campbell has won awards from Poets and Writers, Utah Arts Council, the Fulbright Program, and others. See Johnrobertcampbell.com for samples of his writing, images, and music.

Laird Christensen teaches writing and environmental studies at Green Mountain College, where he directs the graduate program in Resilient and Sustainable Communities. His poems and essays have appeared in a variety of anthologies and magazines, including *Wild Earth*, *Northwest Review*, *Whole Terrain*, *Northern Woodlands*, and *Utne Reader*. His books include *Teaching about Place* and *Teaching North American Environmental Literature*.

Alison Hawthorne Deming is the author of, most recently, *Zoologies: On Animals and the Human Spirit*, as well as eight other books of poetry and nonfiction. She is Agnese Nelms Haury Professor of Environment and Social Justice at the University of Arizona, where she teaches in the creative writing program. She is a 2015 Guggenheim fellow.

John Elder lives in the Green Mountain village of Bristol, Vermont, with his wife, Rita. He retired from Middlebury College after teaching English and environmental studies there for thirty-seven years. While continuing to sugar in the nearby uplands of Starksboro, he has recently begun to write about the landscape and history of Ireland and about their affinities with those of Vermont.

Aaron M. Ellison is the senior research fellow in Ecology at Harvard University. He studies the disintegration and reassembly of ecosystems following natural and anthropogenic disturbances; thinks about the relationship between the Dao and the intermediate disturbance hypothesis; and reflects on the critical and reactionary stance of ecology relative to modernism. On weekends, he works wood.

Jeff Fearnside's writing has appeared in many journals and anthologies, most recently poetry in the *Fourth River*, *Assisi*, and *Clackamas Literary Review*; nonfiction in *Potomac Review*, *ISLE: Interdisciplinary Studies in Literature and Environment*, and *The Chalk Circle: Intercultural Prizewinning Es-*

says (Wyatt-MacKenzie Publishing); and fiction in *Fourteen Hills, Soundings East*, and *Everywhere Stories: Short Fiction from a Small Planet* (Press 53).

Thomas Lowe Fleischner, director of the Natural History Institute and professor of environmental studies at Prescott College, is editor of the anthology *The Way of Natural History* and author of *Singing Stone: A Natural History of the Escalante Canyons, Desert Wetlands*, and numerous articles. A naturalist and conservation biologist, he was the founding president of the Natural History Network.

Tim Fox has worked over the last twenty-five years as an owl researcher, vegetation surveyor, archaeological field crew leader, and writer in the fir and hemlock forests of the central Oregon Cascades, where he lives with his wife and son. You can find more of his writing in *Dark Mountain,* issues 4 and 5, and on his blog: Reciproculture.blogspot.com.

Andrew C. Gottlieb is the reviews editor for the journal *Terrain.org*. His work has appeared in many journals, including the *American Literary Review, American Fiction, Best New Poets, Beloit Fiction Journal, Ecotone, Fly Fish Journal, Poets and Writers, Salon.com, saltfront*, and *Sugar House Review*. He's been writer in residence in a number of wilderness locations. Find him at Andrewcgottlieb.com.

Vicki Graham, author of three collections of poetry, *The Hummingbird's Tongue, The Tenderness of Bees*, and *Alembic,* teaches English, creative writing, and environmental studies at the University of Minnesota, Morris. She divides her time between the Minnesota prairie and the southern coast of Oregon. Her poems and articles have appeared in *Poetry*, the *Midwest Quarterly, ISLE, EarthLines Magazine,* and *Seneca Review,* among others.

Jane Hirshfield's eight poetry books include *The Beauty* (Knopf, 2015) and *Come, Thief* (2011); other works include two essay collections and four books cotranslating the work of world poets of the past. Her work appears in the *New Yorker,* the *Atlantic,* and eight editions of *The Best American Poems*. She is a current chancellor of the Academy of American Poets.

Linda Hogan, a Chickasaw novelist, essayist, and environmentalist, is the author of numerous poetry collections, novels, and nonfiction books. Intimately connected to her political and spiritual concerns, Hogan's work deals with issues such as the environment and ecofeminism, the relocation of Native Americans, and historical narratives, including oral histories. She is the Chickasaw Nation writer in residence.

Freeman House is a former commercial salmon fisherman and cofounder of the Mattole Watershed Salmon Support Group and of the Mattole Restoration Council. The author of the groundbreaking book *Totem Salmon: Life Lessons from Another Species*, he lives in Petrolia in the Mattole River Valley of northwestern California.

Dr. **Robin Wall Kimmerer** is a mother, botanist, writer, and distinguished teaching professor at the SUNY College of Environmental Science and Forestry in Syracuse, New York, where she is the founding director of the Center for Native Peoples and the Environment. Her writings include *Gathering Moss*, which was awarded the John Burroughs Medal in 2005, and *Braiding Sweetgrass: Indigenous Wisdom, Scientific Knowledge and the Teachings of Plants*, published in 2013.

Joan Maloof is a scientist, a writer, and the founder and director of the Old-Growth Forest Network, a nonprofit organization creating a network of forests that will remain forever unlogged and open to the public. Her books are *Teaching the Trees: Lessons from the Forest* and *Among the Ancients: Adventures in the Eastern Old-Growth Forests*. Maloof is a professor emeritus at Salisbury University in Maryland.

Jerry Martien lives in the Humboldt Bay region of Northern California, where, since 1970, he has been a poet in the schools, bookseller, carpenter, public nuisance, and teacher. He is the author of several chapbooks of poetry; a poetry collection, *Pieces in Place*; and *Shell Game: A True Account of Beads and Money in North America*.

Kathleen Dean Moore is a philosopher, writer, and climate advocate, coeditor of *Moral Ground: Ethical Action for a Planet in Peril,* and author of award-winning books of essays: *Riverwalking, Holdfast, Pine Island Paradox,* and *Wild Comfort.* Oregon State University Distinguished Professor of Philosophy Emerita, she is the cofounder of the Spring Creek Project, where she serves now as a senior fellow.

Lori Anderson Moseman runs Stockport Flats, a poetry press she founded in the wake of Federal Disaster no. 1649, a Delaware River flood. Her poetry collections are *Utmost Brevity* (Spyten Dyvil, 2015), *All Steel* (Flim Forum Press, 2012), *Temporary Bunk* (Swank Books, 2006), *Persona* (Swank Books, 2003), and *Cultivating Excess* (Eighth Mountain Press, 1991).

Naturalist **Robert Michael Pyle** writes essays, poetry, and fiction along a tributary of the lower Columbia River in southwest Washington. His eighteen books include *Evolution of the Genus Iris: Poems, Mariposa Road, Chasing Monarchs,* and *Wintergreen.* John Burroughs Medalist, distinguished alumnus of the Yale and Washington forestry schools, and Guggenheim fellow, Pyle founded the Xerces Society for Invertebrate Conservation.

Pattiann Rogers has published fourteen books, most recently *Holy Heathen Rhapsody* (Penguin, 2013). Rogers is the recipient of two grants from the National Endowment for the Arts, a Guggenheim fellowship, and a literary award from the Lannan Foundation. Her poems have received five Pushcart Prizes, made two appearances in *Best American Poetry,* and made five appearances in *Best Spiritual Writing.* Her papers are archived at Texas Tech University. She is the mother of two sons and has three grandsons. She lives with her husband, a retired geophysicist, in Colorado.

Scott Russell Sanders is the author of twenty books of fiction and nonfiction, including *A Private History of Awe* and *A Conservationist Manifesto.* The best of his essays from the past thirty years, plus nine new essays, are collected in *Earth Works,* published in 2012 by Indiana University Press. In 2012 he was elected to the American Academy of Arts and Sciences. He is a Distinguished Professor Emeritus of English at Indiana University. More can be found at Scottrussellsanders.com

Scott Slovic is a professor of literature and environment and chair of the English Department at the University of Idaho. The longtime editor of *ISLE*, his recent coedited books include *Currents of the Universal Being: Explorations in the Literature of Energy*, *Ecocriticism of the Global South*, and *Numbers and Nerves: Information, Emotion, and Meaning in a World of Data.*

Michael G. Smith is a chemist and outdoors enthusiast. His poetry has been published in many literary journals and anthologies. *No Small Things* was a finalist for the New Mexico–Arizona Book Award. *The Dippers Do Their Part*, a collection of *haibun* and *katagami* coauthored with Laura Young, recalling their Spring Creek Project residency at Shotpouch Cabin, will be published in 2015.

Dr. **Tom A. Titus** is a research associate in the Institute of Neuroscience at the University of Oregon, where he teaches a course on amphibians and reptiles of Oregon. He is president of the Eugene Natural History Society, and his creative writing includes the memoir *Blackberries in July: A Forager's Field Guide to Inner Peace.*

Brian Turner is one of New Zealand's best-known poets and nonfiction writers. His numerous awards include the New Zealand Book Awards for Poetry in 1993 and 2010 and the Lauris Edmond Memorial Award for Poetry in 2009. He was the Te Mata Estate New Zealand Poet Laureate 2003–2005 and received the New Zealand Prime Minister's Award for Literary Achievement in Poetry in 2009.

Bill Yake, now living among the fir and redcedar forests bordering the Salish Sea, was born, raised, and first educated where eastern Washington pine forests grade into the remnant black hawthorn swales and eyebrows of the Palouse Hills. Bill's poems can be found collected in *This Old Riddle: Cormorants and Rain* and *Unfurl, Kite, and Veer* (Radiolarian Press, Astoria).

Maya Jewell Zeller grew up in rural areas of the Pacific Northwest. She is the author of *Rust Fish*, a collection of poetry from Lost Horse Press; her nonfiction and poetry appear in *Pleiades*, *West Branch*, *Cincinnati Review*, and elsewhere. Maya lives in Spokane with her husband and two small children.

ABOUT THE PHOTOGRAPHER

Writer and photographer **Bob Keefer** spent four decades as a newspaper writer before retiring in 2013 to pursue independent writing and photography. In 2006 he was a fellow at the National Endowment for the Arts' Journalism Institute for Theater and Musical Theater in Los Angeles. Keefer has shown his hand-colored photography at the Karin Clarke Gallery and the Jacobs Gallery in Eugene and at the Blue Sky Gallery in Portland.

Acknowledgments

Thanks to all the writers, philosophers, and artists who have spent time at the H. J. Andrews Experimental Forest as part of the Long-Term Ecological Reflections program and who contributed works to the program's journal, the *Forest Log*.

The U.S. Forest Service's Pacific Northwest Research Station has been a vital supporter of the Reflections program from its inception. We especially thank research leaders Jim Sedell, John Lawrence, Cindy Miner, and Rob Mangold.

The foundation and inspiration for the Reflections program has been the Long-Term Ecological Research program at the Andrews Forest, supported by the National Science Foundation, U.S. Forest Service Pacific Northwest Research Station, Oregon State University, and the Willamette National Forest.

We extend our gratitude to the managers of the Andrews Forest headquarters site, notably Kathy Keable, Mark Schulze, and Kathleen Turnley, who have been excellent hosts to the many visiting writers and artists. Thanks also to the scientists, including Steve Ackers, Mark Harmon, Jay Sexton, and others, who have shared their knowledge of the forest with visiting writers.

Our efforts have been buoyed and inspired by the vigorous, diverse Reflections-like programs at many sister Long-Term Ecological Research sites and other places of long-term inquiry in the natural world, described in part at the Ecologicalreflections.com website.

Thanks also to all the good people at the University of Washington Press, especially our editor, Regan Huff, for her enthusiasm and thoughtful guidance.

And finally, we are grateful to the steady leadership of the Spring Creek Project for Ideas, Nature, and the Written Word, based in the School of History, Philosophy, and Religion at Oregon State University, and especially to Spring Creek's visionary cofounders, Franz Dolp and Kathleen Dean Moore.

PREVIOUSLY PUBLISHED WORKS

"Scope" by John Campbell was first published in *Terrain.org.*

"The Other Side of the Clear-Cut" by Laird Christensen was originally published (in different form) as "A Tree Falls in the Forest" in *Whole Terrain,* 2012.

"The Owl, Spotted" by Alison Hawthorne Deming was first published in *OnEarth* (28, no. 3 [Fall 2006]).

"The Web" by Alison Hawthorne Deming was first published by *Orion* (26, no. 2 [March–April 2007]). Both "The Web" and "Specimens Collected at the Clear-Cut" were subsequently included in the author's *Rope* (Penguin Press, 2009).

"Portrait: Parsing My Wife as Lookout Creek" by Andrew C. Gottlieb first appeared in *Best New Poets* (2013).

"For the *Lobaria, Usnea,* Witch's Hair, Map Lichen, Ground Lichen, Shield Lichen" by Jane Hirshfield first appeared in the *Atlantic* (February 24, 2011) and was subsequently included in the author's *Come, Thief: Poems* (Knopf, 2013).

"Wild Ginger" © by Jane Hirshfield first appeared in *Orion* (29, no. 6 [November–December 2010]: 80).

"Listening to Water" by Robin Wall Kimmerer first appeared (as "Witness to the Rain") in *The Way of Natural History* (ed. Thomas Fleischner, Trinity University Press, 2011).

"Hope Tour, Three Stops" by Lori Anderson Moseman first appeared in *Trickhouse* (Winter 2010) and was subsequently published (under the title "Increment Borer | Comparative Analysis") in the author's *All Steel* (Flim Forum Press).

"In the Experimental Forest" by Robert Michael Pyle was first published in the author's *Evolution of the Genus Iris* (Lost Horse Press, 2014).

"The Long Haul" by Robert Michael Pyle was first published in *Orion* (23, no. 5 [September–October 2004]); it was subsequently collected in the author's *Tangled Bank: Writings from Orion* (Oregon State University Press, 2012).

"Genesis: Primeval Rivers and Forests" by Pattiann Rogers was first published in *Georgia Review* (60, no. 2 [Summer 2006]).

"This Day, Tomorrow, and the Next" by Pattiann Rogers was first published in *OnEarth* (29, no. 2 [Summer 2007]). This poem and "Genesis: Primeval Rivers and Forests" were subsequently collected in the author's *Wayfare* (Penguin, 2008).

"Mind in the Forest" © by Scott Russell Sanders was first published in *Orion* (28, no. 6 (November–December 2009): 48–53; it was subsequently collected in the author's *Earth Works: Selected Essays* (Indiana University Press, 2012), 333–343.

"Slough, Decay, and the Odor of Soil" by Bill Yake was first published in *Windfall: A Journal of Poetry of Place* (Spring 2011).